Studies in Logic
Logic and Argumentation
Volume 103

The Fertile Debate
Affective Exploration of a Controversy

Studies in Logic Series Editor
Dov Gabbay   dov.gabbay@kcl.ac.uk

The Fertile Debate

Affective Exploration of a Controversy

Claire Polo

ISBN 978-1-84890-437-8

College Publications
Scientific Director: Dov Gabbay
Managing Director: Jane Spurr

http://www.collegepublications.co.uk

TABLE OF CONTENTS

Acknowledgements

I would like to thank all the people who have contributed to the pedagogical activities and linguistic research that I report in this book. First of all, I would like to thank the trio of Kristine Lund, Gerald Niccolai, and Christian Plantin, who have accompanied me throughout my doctoral work, guiding me and showing me enough confidence to help me develop an autonomous thought. My thanks also go to all those who took part in one way or another in the realization of this fabulous body of research: students, facilitators, teachers, the network of the *Petits Débrouillards*, Zeynab Badreddine, who participated in video-recording the sessions in France and Mexico. I also want to thank the different scientific communities that welcomed my work despite its interdisciplinary nature, and its tendency to evolve and develop on the margins of several research traditions. You allowed me, through exciting exchanges, to enrich it well beyond the version of my thesis defended in April 2014. Thank you to John Woods, for trusting in the value of my work for his *Studies in Logic and Argumentation*, and for his patience during the translation process that took longer than we initially thought. He gave me the opportunity to make this work accessible to a non French-speaking audience of researchers and practitioners worldwide.

Finally, I would like to thank the youngest people around me who inceasingly give me food for thought about our individual and collective ways of thinking about our world and its future: my students, my nephews, and my son Enki. He is beginning to argue, defining himself sometimes as "big" sometimes as "small", perfectly distinguishing between the paths that these qualifiers allow him to take.

Preface by Christian Plantin

Claire Polo is interested in socioscientific controversies (now SSC); this particularly complex research object began to be seen as a coherent whole in the late 1990, see Claire Polo's introduction on p. 10; or, in French, Pablo Jensen's invitation for researchers and citizens to meet during scientific *cafés* (1998). Such questions can be presented and discussed in various ways, at school or elsewhere, with or without relating to the curriculum. Discussions can take place in a relatively informal way, or be arranged in different perspectives, ranging from the acquisition of specific scientific knowledge, or general scientific education, up to political and civic education.

A research community exists around such objects, which, to be apprehended correctly, require using and coordinating knowledge and theories coming from different disciplines, not necessarily familiar with dialoguing together on specific problems. Claire Polo shows concretely that such an articulation is possible, that it is fertile, and that it should definitely be considered as a research area in the full sense of the term.

In practice, to become research objects, socio-scientific issues and debates need first to be apprehended systematically through a format robust enough to be reproduced in different social and cultural conditions: this is precisely what Claire Polo does when she designs and implements the You Talk to collect data in the United States, Mexico and France. To systematically and not only anecdotally document SSC it requires such a practical expertise and the considerable associated fieldwork.

Analyzing such data then also requires linguistic skills, specifically in interactional linguistics and argumentation. Two remarkable contributions of Claire Polo have to be emphasized: the first one on a few key issues relating to argumentation, the second one about emotions.

 Claire Polo uses a dialogic model of argument, the only way to account for the redoubtable data consisting of students' debates. Such a method is now well established, but its implementation in this book is remarkable in several ways. First, from a methodological point of view, she uses a concept of *discourse object* inspired by Grize. In the discourse developed around an argumentative question, the key terms of the discussion are conceptualized in different ways by the participants, and it is quite fascinating to follow the construction of the concept 'water' in the students' speeches, either categorical or hesitant. More specifically, the use of *lexicometric methods* to define concepts opposed in an argumentative situation is particularly interesting, and provides the conclusions with an objective basis.

Second, regarding the content, the originality of the approach relies in the kind of dialogues studied: a dialogue that must take into account scientific data, and that tends to adjusting and the constructing knowledge.

Traditionally, since Aristotle until Pereleman & Tyteca, argumentation is opposed to demonstration, "those who prove to those who argue". The theory of argumentation within language fits into this tradition, even strengthening such opposition:

> Not only the words do not allow for demonstrating,
> but they also little allow for this degraded form of
> demonstration that arguing would be.
> [Argumentation] is only a dream of discourse
> (Ducrot, 1993, p. 234)[1].

On both sides of the divide, there is a total consensus. On the epistemological side, the gap between the "two cultures" was radicalized by Gaston Bachelard, twenty years before Perelman, by opposing science and opinion, which is "always wrong" (Bachelard, 1938 , p. 14). Spontaneous knowledge would focus on interest and use (ibid.), which would hinder the objectivity that characterizes real knowledge carried by the scientist. In contrast, opinions are generated and expressed in natural language, and attached to subjects, precisely constituted by exercising such language (Benveniste, 1966, p. 260). Against Bachelard, one can argue, as Claude Lévi-Strauss, that natural language and interests, properly "arranged" can produce real knowledge.

This book challenges on a concrete case, this foundational rupture between evidence and argument. It belongs to a growing research movement, that studies argumentation as an instrument for knowledge building. After all, science teachers, *willy-nilly* , have to adapt to their students' early beliefs and make them evolve, all this through natural language, "until there is a clear gain in departing from it", as says Willard van Orman Quine about elementary logic (1980, p. x).

Thus, Claire Polo's work contributes to the construction of a common language between what C. P. Snow, in his famous 1959 lecture, called "the two cultures", namely the sciences and the humanities. As soon as argumentation aims at apprehending human problems, it must take into account the fact that, in today's world, these human problems also

---

[1] Own translation from the French: "Non seulement les mots ne permettent pas la démonstration, mais ils permettent aussi peu cette forme dégradée de la démonstration que serait l'argumentation. Celle-ci n'est qu'un rêve du discours" (Ducrot, 1993, p.234).

depend on science and technology. Argumentation then cannot limit to common sense and preconceptions; it has to define a new relation to that 'other culture', the scientific and technical culture. There is no more argumentation without information – if there ever was.

The treatment of emotions offered by Claire Polo is particularly remarkable. Just as, on its right, the field of argumentation was built in opposition to demonstration, on its left, it was built in opposition to emotions. All well-known rhetorics take into account the question of passions, – except for the " New Rhetoric" of Perelman & Olbrechts-Tyteca, who declare preferring the term *value* to *passion* , considered" less pejorative" ([1958] , p. 630).

However, Claire Polo's book analyzes the expression of emotions, not in respect of a specific rhetoric, or as an emotional component to include to humanizing thing a little bit as stated by Michel Houellebecq, but as something inherent to both the *situation* of debating and the *theme* of the debate.

On the one hand, the argumentative situation when not limited to a brilliant contest between professionals of paradox, is emotional by definition. Besides engaging their faces, participants then see their identity beliefs called into question, and are challenged by facing other beliefs and practices that may directly contradict their moral principles and habits. Participants to a serious argumentative situation experience strange otherness.

Moreover, it is clear that, in the current climate context, a question like that of 'water' can be seen from the perspective of a possible *shortage*, a word that expresses indirectly (by inference) a sense of *concern* or *anguish*. But it can also be apprehended as a contemporary expression of the many challenges that humanity has faced and overcome in its development; and this second vision of the question as a *challenge* carries *a positive feeling of excitement* that can be *joy*.

This is precisely what Claire Polo shows about the emotional framing of the object under discussion. In particular, this study includes the parameters of *distance in space* ("this issue concerns me and my relatives" vs "concerns people far away") and *time* ("it is urgent" vs "for the moment everything is still okay"), and, of course, the question of *control* , which plays an essential role in defining emotional attitudes. In short, to put it in students' terms, "we will run out of water, the poor will suffer" vs "right now, we're a little under pressure but we will adapt". Such utterances are inseparably emotional–cognitive constructs.

Needless to insist on the social and educational value of the discussions on which Claire Polo provides a landmark study. They are of public interest, and should be supported as a great national cause, in an era when *fake news* and conspiracy theories circulate, quickly spreading and reinforcing on social networks, when people pretend to affirm their emotional identities by claiming their freedom to say whatever comes into their head, following models from the higher spheres. This is not by chance that new forms of negationism express in a violent language; what is at stake is to intimidate the most humble opponent, who would remember, for instance, that even in natural language, 2 + 2 often equal 4.

References of the Preface

Bachelard, G. (1938). *La formation de l'esprit scientifique*. Paris : PUL.

Benveniste, É. (1966). De la subjectivité dans le langage. *Problèmes de linguistique générale, I*. Paris: Gallimard.

Ducrot, O. (1993). Les topoi dans la « Théorie de l'argumentation dans la langue. In C. Plantin (dir.), *Lieux communs, topoi, stéréotypes, clichés* (p. 233–248). Paris: Kimé.

Lévi-Strauss, C. (1962). *La pensée sauvage*. Paris : Plon.

Perelman, C. et Olbrechts-Tyteca, L. (1976) [1958]. *Traité de l'argumentation. La nouvelle rhétorique*. Bruxelles : Éditions de l'Université de Bruxelles.

Quine, W. v. O. (1980). *Elementary Logic. Revised edition*. Cambridge, MA. & London, England: Harvard UP.

Introduction: *affective* exploration of a controversy

Any debate, especially when dealing with 'hot', socially controversial topic, may turn 'sterile', and get blocked on unceasing repetition of fighting opinions. Participants then oppose to one another without understanding each other, confront viewpoints without enriching them. Among the factors responsible for lowering the quality of the debate, emotions, identity dynamics, leadership effects, ignorance are often mentioned. Does it mean that collective intelligence is impossible in argumentative practice about socially accute question? I do not believe so, even if, when it comes to apprehending a sociosicentific controversy (here after SSC), a socially accute question with high technico-scientific content, the exercize may be even more difficult. Exploring such a problem cannot be achieved on the basis of a unique type of expertise, and it can only proceed through dialogue. On the plane of ideas, it involves a dialogue between disciplines, between reference and experiential knowledge, between knowledge of any kind and emotions, norms and values guiding our choices as human beings and as societies. On the concrete plane, putting into debate SSC has also been vindicated as a democratic necessity (Callon, Lascoumes, Barthe, 2001) and experimented through a pedagogy of dialogue[1]. Today, such trend leads to the multiplication of discussion tools fostering citizens' participation to decisions or at least to the elaboration of recommendations about such issues. Non-profit organizations and foundations of popular education are playing a great role in this process. For tomorrow, the challenge is to train children and teenagers to critically but constructively adress complex problems such as genetic engineering progress, nanotechnological potentialities, global warming, exhaustion of natural resources... Here again, a dialogic approach is developping, notably — but not exclusively — in schools, aiming the improvement of comunicative and argumentative skills necessary to reason about such issues. Still, in the classroom like anywhere, you do not always reach truly cognitively enriching interactions, sources of learning, only by opening a discussion space. What characterizes, as opposed to the 'sterile

---

[1] The fundamentals of dialogic teaching are well presented by Alexander (2017). Today's most active research group on dialogic pedagogy worldwide is the CEDiR (Cambridge Educational Dialogue Reseach : https://www.educ.cam.ac.uk/research/groups/cedir/). Still, within EARLI - European Association for Research on Learning and Instruction, the special interest group 26, Argumentation, Dialogue and Reasoning also addresses the pedagogy of dialogue, from an interdisciplinary perspective integrating argumentative studies: https://www.earli.org/node/49.

debate', a fertile one, a debate that would allow for deep exploration of a controversy and feed the collective construction of stronger arguments than those that each individual might have imagined on his or her own? What are the necessary conditions for such a debate to occur? How to foster its emergence?

I have been interested in such challenges for a long time, first as a citizen, with a sensitivity to the public space trained by my five years in the Institute of Political Studies of Grenoble. I then apprehended them in the *Petits Debrouillards* non-profit organization, both as a pedagogical designer for debate tools about SSC, and as a facilitator of such debates. Later on, I also had to deal with fertilizing discussions as a high school and university teacher. Finally, I have been addressing these challenges as a researcher for one decade, with an interdisciplinary approach at the crossroads of educational and language sciences. In particular, I worked for my PhD in the ICAR laboratory, between 2010 and 2014, on teenagers' argumentative practices debating about drinking water management in 4 schools of 3 countries. I then participated in creating an innovative discussion tool, the YouTalk, adapted and implemented to Mexico, the USA and France. Videorecording of 10 sessions let me deeply analyze students' argumentation in these contexts. This book does not only deliver the main results of this doctoral work, but also share pedagogical tools, methodological and conceptual advances developped during this project and beyond. I hope that anyone interested in exploring a SSC, designing a fertile debate, or studying argumentative interactions will find here food for thought and directions for action.

On the research plane, my very descriptive work consisted in deeply understanding students' argumentative practices, using any concept and method that I found fruitful to characterize the observed phenomena, beyond disciplinary borders. If my work is mostly based on linguistic methods for argumentative interaction analysis, I actually put it into perspective with some studies from intercultural communication, collaborative learning, didactics of science and didactics of socially accute questions, instructional design. I have always hoped that my research could later on be useful for practitioners putting into debate SSC. Understanding how young people argue in such a setting indeed lets us learn about how we can possibly guide them to a constructive exploration of complex issues, to aquiring individual and collective reasoning methods favoring cognitive advancement. I here present the main contributions of this research in these respects. In terms of scientific significance, it is the first study of this size, in an ecological context, gathering data in diverse languages and cultures, which apprehends dialogic reasoning integrating its emotional and relational dimensions. As a whole, this empirical work shows that there is no argumentation but socioaffective. In terms of practical signifance, what can teachers, educators and practitioners of participatory democracy learn from this work? This book aims at emphasizing practical and conceptual tools that may nurture their work. Actually, my approach can be globally understood as an instance of linguistics applieds to profesional practices, in this case, educational ones. Instead of pretending to release a miraculous recipe just to apply, and without underestimating the theoretical implications of the pedagogical tool and practices described, I propose directions for designing and implementing constructive argumentative and collective work settings about complex, controversial issues.

Before getting into the details of the pedagogical setting and studied corpus (chapter 1), I want to emphasize what is to me the major contribution of this book, which its subtitle refers to. What does "*affective* exploration of a controversy" mean? Why not just mentioning 'the place of emotions in the exploration of a controversy'? Indeed, this study is not limited to the use of emotions, but encompasses all the argumentative resources used by the students, also including knowledge, values and norms (chapter 2). Similarly, if such work reveals the emotional dimension of what Grize calls the 'schematization' of discourse objects (1997), this process of discursively creating an image of them is not only emotional. If the way the students talk about the discourse objects influences the debate in favor of their opinion, such argumentative orientation frames their speech both and inseparably affectively and cognitively, through conceptual networking. Chapter 4 precisely describes, converging the results of several methods, the different lights cast on the object 'water' in these debates using analogies, specific themes and relating water to other objects. No need to explicitly mention emotions to conduct these analyses in chapter 4. In turn, I adress emotions when studying group dynamics and the conditions necessary to engage in collective reasoning (chapter 3). But then I am considering the social functions of emotions, rather than the affective logics described upper. In chapter 3, what is at stake is the matter of social recognition in the argumentative interaction and the diverse strategies used to preserve one's own and others' faces, strategies which proves more or less beneficial or detrimental to collaboration. At the end of the day, emotions are everywhere: they are both associated to the objects and to the subjects, and they are inextricably participatinf to the ongoing sociocognitive processes. Consequently, it would be erroneous to consider emotions as external independent variables likely to affect sociocognitive reasoning; and it would be simplistic to describe them as a part or component of argumentation.

To sum up, I use the adjective 'affective' to qualify what exploring a controversy means for three reasons. First, the exploration of the controversy is *affective* because the depicted objects are partly made of affective material, as a canoe can be made of wood. Such exploration is also *affective* since emotion fully soaks it, as if reasoning were literally put into emotion like a sponge in water. Last but not least,

3

the exploration of the controversy is *affective* in the sense that the emotional climate surrounds and penetrates the debate, unceasingly built and renewed through argumentative exchanges. Both emerging from the interactions and structuring them back, the emotional climate consists of three things: the affective tonality confered to the discussed question, the feelings displayed by participants about the preservation of their face, and the level of individual and collective investment in the debate, resulting from the two first components. If the exchanges are safe enough for their faces, still presenting the problem as consistent enough, individuals can engage in exploring a zone of doubt or disagreement. When aligning on such an emotional climate, the latter constitutes a strong common *ground*[2] for collective reasoning. At the end of the day, it is necessary to understand that we cannot explore a controversy but affectively, emotions being raw material and breeding ground of argumentation, to fertilize the debate. Taking into account this affective nature of the exploration of controversies opens fruitful interpretative directions for a research trying to make sense of authentic data differing from canonical models of argumentation. I hope that recognizing such omnipresence of emotions in reasoning will also enrich the practice to imagine more and more fertile debates.

---

[2] The 'grounding' metaphor coming from linguistics, is now commonplace in research about collaborative learning. *Grounding* is the process of establishing or strenthening a commun view necessary for communication and *a fortiori* collaboration. Such *common ground* created in interaction and featuring it consists of the perception of the situation and the task and, more largely, a minimal mutual understanding and shared knowledge and beliefs. For a recent reference publication in linguistics, see Clark, H. H & Brennan, S. (1991). « Grounding in communication ». In L.B. Resnick, J.M. Levine & S.D. Teasley (Eds.) *Perspectives on Socially Shared Cognition*, p. 127–149. Washington DC: American Psychological Association. For an application to the field of collaborative learning, see for instance Baker, M., Hansen, T., Joiner, R., & Traum, D. (1999). « The role of grounding in collaborative learning tasks ». In P. Dillenbourg (red.), *Collaborative Learning: Cognitive and Computational Approaches* (s. 31–63; p. 223–225). Elsevier.

I. An education system for socio-scientific controversies in three countries

Since the end of World War II, the evolution of the economy and scientific research, as well as the rise of the ecological crisis, have favored the emergence of socioscientific controversies (hereafter SSC) in the public space. These societal issues involving scientific and technical aspects can be difficult for ordinary citizens to apprehend. However, the quality of our democracy depends on the extent to which we ensure that everyone can form a considered opinion about such SSC. Schools have a key role to play in training future adults to understand these issues. However, understanding them does not only presuppose the mastery of knowledge but also the ability to put them into perspective with other structuring elements of the public debate, such as the values and material interests of the different stakeholders, or the emotions aroused by the issue under discussion.

The *YouTalk* device, designed thanks to a partnership between the interdisciplinary research laboratory ICAR[3] and the non-profit organization of informal science education *Les Petits Débrouillards*[4] , seeks to contribute to this educational challenge by bringing 12 to 15 year-old students to debate about a socioscientific controversy, guided by a pair of older, volunteer students aged 16 to 19 years old. It was implemented in several countries (France, Mexico, the United States, Brazil[5]), in more or less privileged or underprivileged schools, to reason about drinking water management[6] . I myself organized and filmed 22 YouTalk sequences, and I studied ten of them in detail, carried out in Mexico, the United States and France, for my PhD in Educational Sciences.

---

[3]Located at the *École Normale Supérieure* in Lyon, this research unit (UMR 5191), whose acronym stands for *Interactions, Corpus, Apprentissages, Représentations,* crosses didactics and linguistics, and has historically been one of the first places to develop research on classroom interactions.

[4]The *Petits Débrouillards* network is well-known for its action in the field of scientific and technical education, and is currently one of the most important non-profit organization of environmental education in France. The partnership with the *ICAR* laboratory was set up with the *Association Rhône-Alpes des Petits Débrouillards* (the branch of the Lyon region), that no longer exists.

[5]Five more took place in two schools in Belo Horizonte, Brazil, in the state of Minas Gerais, in May 2014, after my dissertation defense, using a similar protocol.

[6]In France, through the partnership with the *Petits Débrouillards*, three other themes were developped: the life cycle of products and waste management, energy policies and urban ecology.

1.1 The matter of teaching socioscientific controversies (SSC)

A quick historical review of their emergence in the public space allows us to identify the specificities of the didactics of socioscientific controversies (hereafter SSC), which I detail next.

1.1.1 Emergence of SSC in the public space

During the Cold War, technological innovation was considered as a strategic resource. From the 1980s onwards, it also became a key factor of competitiveness in the globalized world economy. In this context, a model of scientific research strongly linked to technical applications was encouraged, having a rapid impact on societies. Some voices raised against the fact that dramatical social choices were then made without democratic debate, and called for the establishment of a "technical democracy" (Callon, Lascoumes, & Barthe, 2001). Moreover, the social studies of sciences, with meticulous analyses of the concrete conditions of the production of research results (Latour & Woolgar, 1979), started to show that subjective choices are operated during the process of the construction of scientific knwoledge. Such approaches contributed to the "desacralization" of science as an objective activity of understanding the world. What some have called the "science–society relationship" also evolved with the emergence of the new paradigm of sustainable development[7] . The awareness of the catastrophic ecological effects of symbolic technologies such as the atomic bomb or pesticides newed the questioning of the applications of scientific activity in society. Moreover, the "fourth pillar" of sustainable development, alongside the economic, social and environmental components, is also starting to be reinvindicated in the public debate: the issues of governance associated with the necessary transition into a new mode of development.

This demand for citizen participation to technological choices has led to the experimentation of new forms of contributions to the public debate such as consensus conferences, citizen forum, dialogic workshops, deliberative polls, interactive technology assessments, controversy exploration seminars, citizen panels, etc. (Callon et al., 2001). One of the challenges of this type of practices is to enable citizens to make "informed" recommendations, as debating on SSC

---

[7]Since the 1990s, "sustainable development" has become an international prism for thinking about the organization and future of societies. The Brundtland report, which marks this emergence, defines it as follows: "Sustainable development is development that meets the needs of the present without compromising the ability of future generations to meet their own needs." (Brundtland, 1987).

supposes basic scientific knowledge. Indeed, both the diagnosis of the problems and the design of possible solutions engage scientific and technical issues. Thus, scientific and technological progress are also presented as one of the factors allowing the shift into a more environmentally friendly development model. However, even with a technical approach of benefits/costs as embodied in the precautionary principle, in a context of uncertainty, decisions cannot only rely on stabilized knowledge. Actors then also make their choices based on their material and "immaterial" interests (moral or political values).

Although questions related to the relations between science, technology and society have been discussed in the public space since the end of the 1960s, often on environmental issues, the problem of teaching such themes, particularly in the science classroom, only developed at the end of the 1980s and the beginning of the 1990s (Bybee, 1987, Aikenhead, 1992). This approach evolved with the emergence of the "socioscientific" movement in the early 2000s, claiming for a more interdisciplinary approach (e.g., Zeidler, Sadler, Simmons, & Howes, 2005), and including concerns of civic education. Thus, "training people who, informed about the research methods, their applications and possible consequences, are able to make informed decisions when the facts are uncertain and to participate in the [corresponding] debates" (own translation from Simonneaux, 2003, p. 189) is becoming a key issue. This concern is institutionalized by policy makers in education, both as the national level (for instance in the French *Socle commun de connaissances et de compétences –* MENESR, 2006), and the international level (for example in the European Commission's *White Paper on Education and Training*, published in 1995). If the school appears to be the key institution for teaching such issues, inviting SSC at school is difficult, and does not go without questioning the teaching practices in depth (Albe, 2009a). On this point, educational policies and the problems raised by the practitioners meet internationally recognized research issues. The literature in science education mentions the need to guide students in their understanding of SSC (e. g. Aikenhead, 1992; Osborne, Erduran, & Simon, 2004; Albe & Simonneaux, 2002 ; Simonneaux, 2003; Zeidler et al. 2005), as well as to have them argue in the science classroom and to train them to deal with controversies (Driver et al., 2000 ; Newton et al., 1999). The criteria and factors of debate quality are of interest to both the science education research and the community of collaborative learning (e. g. Albe, 2006; Désautels & Larochelle, 1998; Driver, Leach, Millar, & Scott, 1996; Lewis & Leach, 2006; Mercer, 1996; Sandoval, 2005). Finally, debates between students on SSC offer an appropriate field of study for the linguistic analysis of argumentation in verbal interactions. I present these concepts regarding these last two fields later on, together with the analyses they allow, with relevant examples drawn from the *YouTalk* corpus. I here only mention the main contributions of research in didactics necessary to specify the matter of teaching SSC.

1.1.2 The specificities of SSC

The Anglo-Saxon science education rather call this type of problems "*socio-scientific issues*", whereas the French tradition of "didactics of socially accute questions in science" or of "socioscientific controversies" puts more emphasis on their controversial dimension. I choose to keep the latter term because it has the advantage of highlighting the fact that science is a social field that is not independent from the rest of society. SSC have four fundamental characteristics that distinguish them from traditional school content knwoledge, and constitute as many challenges for teaching them.

- *Transdisciplinarity*

First of all, SSC are by nature transdisciplinary, whereas school is still largely segmented into disciplines, favoring their treatment within the science classroom. In Anglo-Saxon science education, debates on SSC have been considered, for the last twenty years, as good means to develop argumentative skills necessary to science learning, and to strenghthen or even aKQuire scientific knowledge (Driver et al., 2000; Newton et al., 1999; Oulton et al., 2004; Sadler & Zeidler, 2005). But, if science teachers sometimes try to set up exercises on SSC, they generally feel incompetent to take into account issues that go beyond their discipline (Newton et al., 1999, p. 565). Moreover, in the science classroom, argumentation is generally used in problem-solving situations, aiming at finding the "right" answer, as students are led to "reconstruct" stabilized scientific knowledge. However, by definition, SSC cannot be solved by the use of scientific reasoning alone, which is only one dimension of the problem: "Faced with such questions (...) science is not able to provide a single, definitive answer that would resolve the controversy." (own translation from Albe, 2009b, p. 49). Dealing with SSC in the science classroom disrupts its common perception as a place dedicated to transmission of objective, unquestionable knowledge. In this context, collaboration between teachers from different disciplines is necessary to address these types of issues (Simonneaux & Simonneaux, 2005). Of course, this mission of the school cannot be limited to the science classroom, and should be explicitly addressed in other areas of school life.

The trandisciplinary approach of "didactics of socially acute questions" seeks to build specific didactics, adapted to the very nature of SSC. The idea is to fight against classroom customs that do not favor their in-depth exploration. Indeed, the SSC are often presented in school in a "neutralized" form, corresponding to the "4Rs" of classical school knowledge: as Results, they are supposed to be Realistic, must constitute consensual Referents and are characterized by a Refusal of politics (Audigier, 1999). As a result, "there is a great temptation (...) to 'cool down' the teaching of socially acute questions at school, and thereby to weaken their meaning for the students..." (Legardez, 2006, p. 28). However, if the students should "understand the scientific content involved, its epistemology, identify the controversies about it, and analyze the social consequences of decisions based on science", it is in order to actively participate in the political debate, to "develop an informed opinion about such issues, [become] capable of making choices about prevention, action, use, and of debating about them" (own translation from Simonneaux & Simonneaux, 2005, p. 83).

- *Subjectivity*

Indeed, dealing with a SSC does not imply reaching the only possible rational conclusion, but consists in exploring different ways of apprehending it, involving different views of the world. This is another essential feature of SSC, namely the fact that one can only reason about them *subjectively* (Oulton, Dillon, Grace, 2004). Decisions about SSC involve two aspects: technical-scientific elements and political-ethical choices. The latter depend not only on the state of scientific knowledge and technical possibilities, but also on the power relations between diverging interest groups, and on the values and beliefs that they hold. Thus, ideological, moral or political considerations cannot be excluded from such debates:

> We can say that reflective decision-making must be based on knowledge and the assessement of its credibility, but also that decisions on these issues result from interactions between knowledge and values (for example, values allow one to judge the desirability of various potential consequences associated with alternative decisions). Individuals may make different decisions based on the same knowledge because of different values, or different hierarchies of values. (own translation from Albe & Simonneaux, 2002, p. 135).

While basic scientific knowledge is therefore necessary to understand the issues involved, the adquisition of knowledge is not sufficient for understanding a SSC. It also supposed to consider the ethical and moral dimensions that constitute the main points of disagreement about it (Sadler & Zeidler, 2005).

- *Controversy*

The third fundamental characteristic of SSC is their controversial nature. Related to the second, it implies that they allow for multiple answers that are not contrary to reason (Dearden, 1981, p. 38). Thus,

9

Oulton, Dillon and Grace consider that these questions are defined by 4 characteristics: the groups within the society have different positions on them, these positions are based on different information or different interpretations of the same information, these divergent interpretations are linked to the fact that the groups have different worldviews, due to their belonging to different value systems. They draw two implications for the exploration of SSC: they cannot always be resolved by the use of reason, logic, or experimentation, although they are likely to be resolved when more information becomes available (Oulton et al., 2004, p. 412).

As "acute questions", the characteristic of SSC is that they are controversial both in the scientific field and in the society. They are even "three times controversial", because they are controversial in the "reference knowledge" (scientific knowledge or recognized social and professional practices), in the "social knowledge" (social representations, elements of previous school knowledge "exported" into social knowledge), and in the "school knowledge" (Legardez, 2006, p. 19-20).

Some authors emphasize their controversial dimension within reference knowledge, making uncertainty a defining feature of SSC. This leads them to consider that a good treatment of SSC requires weighing up the risks and benefits, assessing the strength of the evidence presented, the correlations interpreted as indices of causality, and, ultimately, concluding on the uncertainty that one is prepared to admit on a given risk (Millar, Osborne, 1998 ; Monk & Osborne, 1997).

- *Knowledge hibridization*

Finally, a fourth characteristic of SSC lies in the fact that they generate a form of knowledge hybridization. More precisely, to build an opinion on them, we use all the forms of knowledge constituting our relationship to the world, whether it be academic knowledge, common sense, daily experience, etc. But this is very rare at school to open moments where these forms of knowledge are invited to dialogue with reference knowledge. Thus, the SSC bring into the classroom "knowledge developed in social groups that are not usually recognized as producers of valid and legitimate knowledge for teaching", which Albe describes as "profane", with reference to etymology, as knowledge produced "outside the temple", outside the places of "knowing knowledge" (2009b, p. 129).

This cannot be ignored at teaching SSC, as these aspects of informal knowledge and values are determining elements in the construction of a sincere opinion on such issues. Simonneaux and Simonneaux found that it is mainly personal and social factors that influence the oral debate conducted in a whole class on biotechnologies (2005). Students express opinions about SSC that are little changed by the aKQuisition of new scientific knowledge (Simonneaux, 1995; Simonneaux &

Bourdon, 1998). Indeed, these opinions are based on their conception of the nature and the articulation of this worlview with the values that they defend. Albe thus notes that "personal experience appears as a means of interpreting and problematizing the controversy, a reference for arguing and justifying a position" (2006, p. 105). Similarly, the results of Sadler and Fowler (2006) lead them to mitigate the place of scientific knowledge in SSC, by considering it as a mediating "bridge" between school and personal experiences.

The debates on SSC thus imply the use of argumentative skills that young people use informally in everyday situations. The challenge is to motivate their expression in a school exercise. This is why it seemed interesting to us to seek the establishment of a more informal climate to encourage the voluntary involvement of young people in such debates, in particular thanks to the collaboration with the *Petits Débrouillards*, by designing the YouTalk pedagogical situation.

Finally, it should be remembered that dealing with SSC in the classroom inevitably entails a work of didactic transposition, "insofar as the main perspective of [its] teaching is the learning of knowledge and practices, whereas in society it is to actually deal with the issue" (Tiberghien, 2009, p. 9).

Although, depending on the research and practice, more or less importance is given to one or other of the four characteristics of SSC defined above (interdisciplinarity, subjectivity, controversial dimension, knowledge hybridization), there is a consensus recognizing how important is each of them. On the contrary, neither the literature, nor the practices converge on the matter of defining the appropriate posture of the teacher when dealing with SSC in a classroom.

- *Which appropriate teaching posture?*

Teachers, in order to defend themselves against potential accusations of indoctrination, tend to consider that they should stick to the facts, presenting a rational, balanced, and neutral view of SSC (Oulton et al., 2004). Part of the literature does actually recommend this neutral stance (e. g. Henderson & Lally, 1988). However, Dewhurst (1992, p. 159) believes that rationality is not a good basis for discussion because it does not allow students to get to the core points of disagreement and really understand why these controversies are so difficult to resolve. Moreover, their controversial and subjective nature casts doubt on the possibility of providing students with a "balanced" or "neutral" view of SSC (Carrington & Troyna, 1988).

Kelly (1986) identifies four types of possible teaching postures. "Exclusive neutrality" means not addressing controversies and sticking to scientific facts. "Exclusive partiality" refers to the intention to lead students to adopt a particular point of view on a controversial issue. "Neutral impartiality" implies that only students debate controversies, while the teacher remains neutral. Finally, the "committed impartiality"

11

model allows the teacher to provide arguments for opposing views, while sharing her/his own opinion with the class for in a transparent way. Oulton and her colleagues advocate the latter posture, prioritizing the development of students' critical thinking skills through an awareness of potential bias (2004, p. 417). Albe (2009a) also favors a similar teaching posture of critical engagement.

Eventually, the treatment of SSC at school implies several changes in the classroom customs, since they affect the cognitive environment (mesogenesis), as well as the way of moving from one epistemic content to another (chronogenesis), or even the "sharing of responsibilities in the didactic transactions" (topogenesis, Sensevy, 2007, p. 20), and in particular the role of the teacher (Albe, 2009b, p.183–186). What is at stake here is the "didactic contract" (Chevallard, 1992). Indeed, the "forms taken by the didactic contract can constrain the dispositions to debate and thus limit the study of such questions" (Albe, 2009b, p. 190). For example, the survival of a meta rule such as "the teacher must ultimately provide the right answer to the question" (Joshua & Dupin, 1993) can be an obstacle to engaging into these new forms of didactic interaction. This observation highlights the interest of a device such as the *YouTalk*, where the debate is set up as an exception in the school routine.

1.2 YouTalk, an innovative discussion tool

The *YouTalk* is based on a reflection on the school transposition of SSC initiated in 2007 in the ADIS-LST team (Learning, Didactics, Interactions, Knowledge – Languages, Sciences and Techniques) of the ICAR laboratory. The initial idea is to take advantage of the dynamics of social debate that characterize the field of informal education, by adapting the model of the science *café* to school. Designed for an adult audience during leisure time, in a relaxing place (generally in the evening, in a bar), it consists in putting in relation an expert on a SSC with the "naive" public and to allow a dialogue between the concerns of the "citizens" and the state of research in the field. This format, which is extra-disciplinary in nature, offers a voluntary audience a space for active debate, drawing on a variety of knowledge (academic or professional knowledge, media, personal anecdotes, daily experience, etc.). But the simple reproduction of this setting at school, in very different contextual conditions (non–voluntary participation, group constituted beforehand, age, spatio–temporal framework) tends to establish a question–and–answer dynamic between the invited person and the students. This prevents the engagement into a real debate in the classroom. Therefore we started working on an larger adaption of this model in order to encourage more discussion. Particular attention was paid to the establishment, in the classroom, of an interactional dynamic that characterizes usual in science *cafés*, namely the cohabitation of two dialogue spaces: the "public" space, and the parallel discussions between the people sitting at the same table.

After some pilot experiences, the *YouTalk* was born in 2011, in collaboration with the *Petits Débrouillards*[8] . I define here its fundamental principles, first by specifying the educational aims, then by describing the pedagogical sequence, and finally by explaining some important aspects for the spirit of the *YouTalk*.

1.2.1 Presentation of the YouTalk

The YouTalk was designed in the context of an international project of civic and environmental education. In this perspective, the theme addressed is a pretext to encourage a critical questioning approach to a SSC. The main objective is not the aKQuisition of knowledge on the topic, even less of concepts of the science curriculum. Still, the YouTalk offers the opportunity to clarify certain definitions and to aKQuire a common basic knowledge on the topic, in order to be able to discuss it.

- *Educational goals*

Although this activity does not claim to contribute to the teaching the curriculum, some educational goals are in line with transversal skills targeted at school. For instance, it corresponds to elements of the first version of the French "Socle de connaissances et de compétences" in the 'scientific culture' and 'social and civic skills' (MENESR, 2006), as well as to points of the renewed, current version of the national curriculum regarding "languages for thinking and communicating" and "the training of the person and the citizen") (MENESR, 2015). The educational goals of the YouTalk are the following:

> To create a climate of mutual trust, respect and listening that allows each student to express himself or herself. This is done by facilitating the oral formulation of ideas, the ability to synthesize, the relevance of interventions and the ability to listen to others.
> To encourage students to distinguish, to help them strengthen their critical thinking skills, the elements of the discussion that :
> – belong to stabilized science: this corresponds to the knowledge and vocabulary brought in to establish a "basis for discussion";

---

[8]Within the framework of the D3 project led by the *Ecole Normale Supérieure* de Lyon and funded by the regional "University of Solidarity and Citizenship" grant.

– fall into ongoing, still controversial science and as such, are uncertain;
– result of a choice, of a political opinion;
– establish a link between some of these elements.
This goal echoes the one stated in the French official instructions, "to have students understand the distinction between facts and verifiable hypotheses on the one hand, and opinions and beliefs on the other hand" (MENESR, 2015).
To train students to get familiar with a new and complex topic by identifying:
– the necessities and constraints to be taken into account to deal with the problem;
– the main points of disagreement in the debate.
To enable them to build a personal opinion on an SSC, at least to feel capable of having one (to identify the arguments for or against a particular position and/or the uncertainties to be resolved in order to take a position).
To train them to work in a group: cooperate, listen to others' arguments, and contribute to the discussion in a relevant way.
To have the students experience speaking clearly in front of an audience and getting more eloquent.

- *Pedagogical sequence*
Designing a pedagogical tool that actually gets the students fulfill the targeted educational goals rather than simply "doing the activity" is an eternal challenge (Jimenez–Aleixandre, Rodriguez, Duschl, 2000). We designed an activity lasting about two hours (100 minutes with a 10-minute break), consisting of five phases: introduction, three thematic phases, conclusion. The whole sequence is led in pairs, one person being in charge of most of the exchanges with the class, and the other one observing what happens with a concern of providing opportunities to participate to every student, notably by supporting the peer interactions in small groups.
The activity is punctuated by a slide show displaying a multiple choice questionnaire (henceforth MKQ). A brief introduction announces the theme and explains the rules of the debate. Particular attention is paid to the distinction between opinion questions (OQ), which have no "right answer" and on which it is likely that not all students will agree, and knowledge questions (KQ), which function more like a quiz, and for which a right answer is identified and justified by reference to a specific source of information. In order to allow the students to become familiar with the electronic voting system and to get to know them better, a few questions are asked (age, gender, attitude towards speaking in class) in the introductory phase of the *café*. The main

14

question (now MQ) is presented, and the initial opinion of each student is registered through an electronic, anonymous survey. The question is:

> In your opinion, in the future, whether a person has access to dinking water will depend on...?
> a) on how rich the person is
> b) on how physically able the person is to live with lower water quality
> c) on efforts made, starting now, to save water by using less and to protect water resources
> d) on where on the globe the person is born
> e) on nature's capacity to adapt to our needs for water
> f) on scientific advances.

The body of the *café* is structured in the form of 3 thematic phases corresponding to 3 OQ allowing to explore the MQ:

> OQ1. In your opinion, which potential sources of drinking water are the most promising for the future?
> a) The discovery of new fresh water deposits
> b) Water that we don't use today (that which, is economized)
> c) Desalination of seawater
> d) Climate change leading to more rain
> e) New techniques for the depollution of water
> f) None of these are promising: water is going to be in short supply and will become THE conflict of the 21st century.
>
> OQ2. Which of these things would you be the most willing to do?
> a) Take fewer baths and showers
> b) Use my phone or computer for a longer time before getting a new one
> c) Eat less meat
> d) Use dry toilets
> e) All of that, and even more!
> f) I'm not ready to make any of those efforts.
>
> OQ3. How should the price of water be determined?
> a) Drinking water should be free
> b) Drinking water should be sold at a price that covers the cost of its production
> c) Drinking water should be sold at a price that depends on its quality
> d) Drinking water should be sold at a price that depends on how it is used

       f) Drinking water must be free up to a reasonable amount, beyond which it should be sold at a high price.

Each thematic phase starts with 3 KQ giving information to help understand the OQ asked at the end of the phase and deal with it. The KQ are followed by a brief discussion "at the table" and an anonymous individual electronic vote. The results are then presented to the class, with the correct answer displayed in green, and a brief informative slide called a "information desk" justifies the expected answer, mentioning the source used. Then, a short exchange in the class takes place based on this slide, to make sure it is understood. The idea is that the KQ on a topic will activate and/or update students' knowledge necessary to reason about the following OQ. As an example, figures 1 and 2 show the slides of the MCQ corresponding to the first KQ of the first thematic phase, the state of water resources, and the corresponding "information desk".

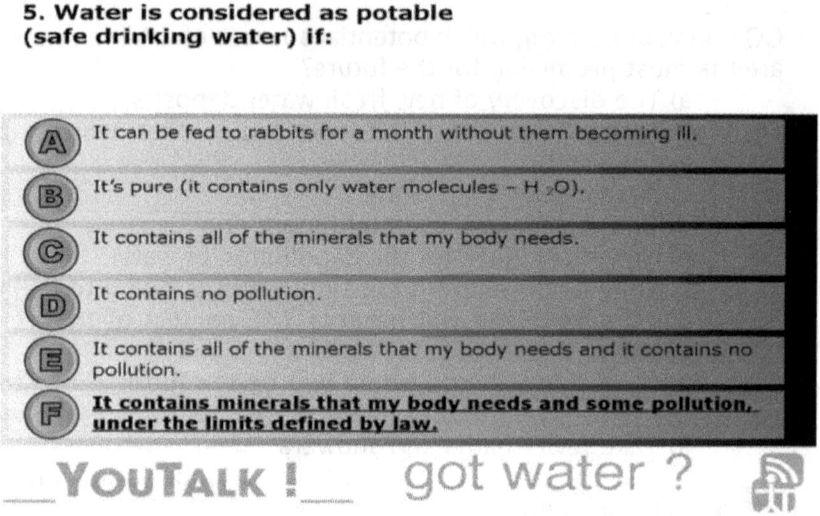

Figure 1. First knowledge question of the *YouTalk* on drinking water.

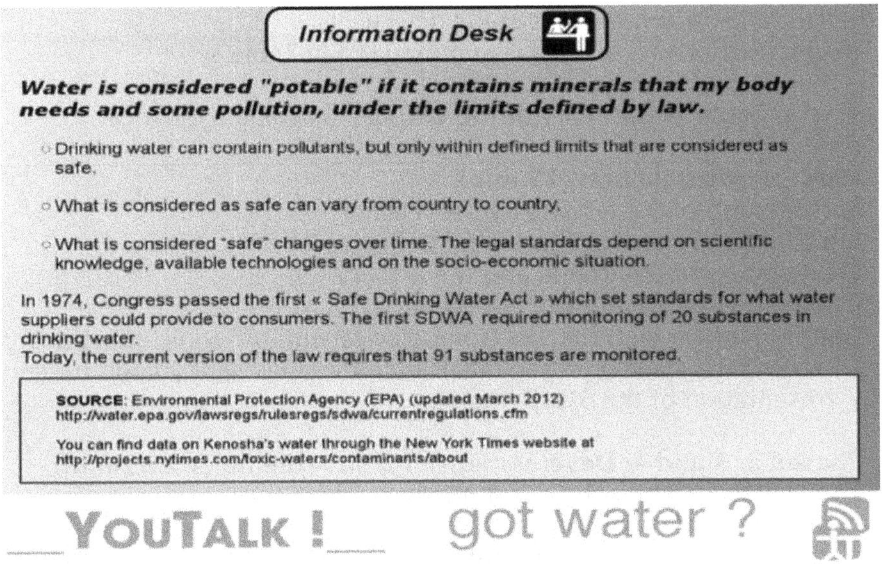

Figure 2. Information desk of the first knowledge question of the *YouTalk* on drinking water.

The OQ (distinguished from the KQ by a color code) are apprehended in another way: the students have more time to discuss them, and must agree by table (of 3 to 5 students) to choose one of the proposed answers. This collective vote is then expressed by the display of a letter corresponding to the answer chosen by the group using a card placed on a pulpit. The panorama of opinions expressed by the different groups in the class is thus immediately visible, which makes it possible to identify the existing divides. The results of this real–time survey can be used to stimulate the group debate that follows ("Three groups think that... Could someone explain why? ... Why do others disagree?) This discussion involves all students (there is no "spokespersons" for the groups, and a student may even express a different opinion from the one displayed at his or her table if he or she wishes). While one facilitator coordinates speech turns, the other one takes notes on the board about the key aspects of the problem to be considered. Finally, the students individually and anonymously vote on the OQ using an electronic system, and the distribution of the answers expressed in the room is quickly displayed on the screen before moving on to the next phase.

During the conclusion, the facilitator who has taken notes makes a quick summary of the debates, then the MQ appears again, and the students answer according to the OQ protocol, with a longer debate time. As for all OQ, it ends with the individual and anonymous vote of the students, using the electronic device.

This instructional sequence is summarized in Table 1.

Table 1. Summary of the *YouTalk* teaching sequence

Phase 1: Introduction (15 min) Presentation of the activity by the teacher Introduction of the facilitators by the researcher Presentation of the theme and the activity Explanation of the rules of the game Getting familiar with the electronic device with off-topic questions – Reading the Main Question (MQ) and individual, anonymous survey – Presentation of the 3 sub-themes **Phases 2, 3 and 4: Development of a sub-theme (3 x 21 min)** Knowledge Question 1 (QC) (4 min) – Reading (30 seconds) – Small group discussion (1 min) – Individual, anonymous vote (30 seconds) – Results of the votes and "information desk " (2 min) QC 2 (idem, 4 min) QC 3 (idem, 4 min) Opinion Question (OQ) (9 minutes) – Reading (30 seconds) – Small group discussion (3 min) – Collective vote by displaying a card with the letter corresponding to the chosen answer – Class debate with notes taken on the board (5 min) – Individual, anonymous voting (30 seconds) **Phase 5: Conclusion: discussion on the MQ (12 min)** – Summary of notes taken during whole class debates (3 min) – Reading the MQ again (30 seconds) – Small group discussion (3 minutes) – Collective vote by displaying a card with the letter corresponding to the chosen answer – Class debate (5 min) – Individual, anonymous vote (30 seconds)

1.2.2 Clarification of the spirit of the activity

Three points need to be clarified in order to fully understand the spirit of the *YouTalk*: the interest of using a MCQ, the survey methods and the role of the facilitators.

- *Using MCQ to foster oral argumentation*

The use of a tool as constraigning as a MCQ may seem antinomic with the informal atmosphere targeted in a junior scientific *café*. But asking students to debate 'freely', without providing them with any support to frame the discussion, does not ensure that they would engage in deep argumentation. Real issues of dominance are at work: the debate can be an opportunity for some students to take power over others, who would find themselves excluded, or relegated to a less beneficial task in terms of potential learning (Muller Mirza, 2008, p. 14). Argumentation puts students at risk of 'losing their face' in front of their peers (Douaire, 2004). The "free" debate then tends to be free only for those who know how to take advantage of it, being used to argueing, or displaying a *leadership* ensuring them access to the conversational floor and discouraging others from participating.

The "framing of the debate" can avoid these perverse effects. Introducing mediation tools in the true sense of the word, encouraging the participation of all with mandatory listening times, for instance, is sometimes necessary for each student to be free to speak out. For example, Albe has pointed out that the introduction of tools to support collaborative discussion improves the quality of debates among students on SSC (2009b, p. 157).

Nevertheless, the choice of an electronic MCQ as a discussion tool may be surprising. The risk is to enter into a simple "quiz" logic in which the students would find themselves trying to "guess" the right answer on a subject they do not master, and would therefore inevitably be put in a situation of failure. This is why explanations are provided to justify the expected answer for the KQ during the "information desks", slides of a few sentences, tables or graphs, citing their sources. The presence of KQ is very important, as they provide information common to all students, which can later nurture the debate on the following OQ and, in turn, on the MQ. The aim is to avoid "exchanges of ignorance" (Clarke, 1992) and to allow the development of quality argumentation thanks to the use of a minimum of knowledge. In class, observations show that the students who have the most knowledge have a deeper reasoning, with less incoherence and contradictions (e. g. Sadler & Zeidler, 2005). This need for knowledge to argue about a SSC was also raised by Lewis and Leach (2006). They showed that a limited understanding of the context and associated science resulted in either no response or an unargued affective response. Their results also point that a more reasoned response is possible on a more familiar topic. However, their findings are mitigated on the effect of providing additional information to prior knowledge. Students seem to be able to reconsider their judgments thanks to the investigator's input, except when the new information is too far from their experience and irrelevant to their immediate lives. This is why we tried to relate the

SSC under consideration to the students' daily lives all along the MCQ[9] . Lewis and Leach's conclusion is interesting: some understanding of the science associated with the SSC is necessary, but it is relatively modest. Sadler and Fowler proposed a threshold model to schematize this relationship knowledge – quality of argument on SSC (2006).

OQ are designed to encourage the exploration of the MQ. In this sense, they are "derived questions" as Plantin defines them as "*questions whose exploration appears necessary in the context of the exploration of the main question*" (Plantin, 1996). Thus, when creating the MCQ, we first looked for at the MQ and then formulated the OQ necessary to explore to build an opinion on the MQ.

The fact of proposing a series of closed answers to the students on a question that merits debate is a very specific aspect of the YouTalk. The MCQ format is very frustrating and the question–answer set proposed is designed to have the students want to "check off" several answers, or to propose other, mitigated own alternatives. However, the students are asked to vote for an answer that do not necessarily suits them completely, either individually or collectively. They have to choose the option they prefer among those proposed. This frustration aims to stimulate the verbalization of disagreements and arguments used to compare these answers and explain how they made their choice. This pedagogical setting aims to get the students express why the chosen answer is more appropriate than the others, but also why such a vote cannot fully express the group's position, or even formulate a new proposition. The propositions made this way contribute to the collective exploration of the MQ, of which the note-taking keeps a record. The fact of restricting the possible answers at the beginning also aims at avoiding a demonization of the opposed position, and engagement into a polarized debate around two caricatural positions. This binary logic has been pointed out as a frequent limitation of debates (Albe, 2009b, p. 157). Here, the students are not divided between a "for" position and a "against" position, but between six possible positions.

- *Real–time opinion polls*

The concern to encourage the circulation of arguments between the private space of debate ("at the table") and the "public" conversation led us to alternate between times of discussion in small groups and times of debate in the whole class, with a real–time poll of students' opinions. In this respect, one polling modality plays a particular role: it is the collective, non–anonymous vote, by table group, at the end of

---

[9]Actually, the debate theme is consistent with the four dimensions to consider in education to SSC: psychological (motivating, close to the students' daily life); cognitive (complex theme, but on which the students have some knowledge); social (controversial) and didactic ("learnable") (Joaquim Dolz, Schneuwly, & De Pietro, 1998).

their discussion on the OQ. This practice fulfils two main functions. On the one hand, it makes it possible to clearly distinguish between KQ and OQ by differentiating how to reason about them. On the other hand, the fact of having to position oneself collectively confers a real stake in asserting one's point of view and submitting it to the criticism of others, or seeking to reconcile it with that of others, during the group discussion. Moreover, it is on the basis of this group survey that the class discussion is stimulated, requiring the construction of strong arguments beforehand. Actually, the literature shows that speaking is easier in small groups for some people (Mercer & Littleton, 2007), particularly students with low academic scores (Simonneaux and Simonneaux, 2005). Alternating between small group and a class discussion aims at minimizing the effects of dominance and "preparing" the confrontation between different views, carried and elaborated collectively. Indeed, the students generally seem to collaborate better in the discussion if they have had a phase of individual or small group preparation beforehand (e.g. Schwarz & Glassner, 2007).

In any case, it is important to consider the fact that the groups may not reach agreement, and it is interesting to keep a step allowing individual and anonymous positioning on the OQ, after the debates, as research data, but also as a survey visible to all the participants, which may serve the facilitation of the activity. Indeed, anonymity favors the expression of a change of opinion after the debate, and the "deviant" positioning with respect to the majority opinion (Ainsworth et al. , 2010). On the other hand, anonymizing the debate favors above all off-topic and off-frame behavior, which is why it is better, in the whole-class debate, to be able to identify which table chose which option. In the pedagogical situation studied, only individual votes are anonymous, which is intended to encourage the sincere participation of students (e. g. Draper & Brown, 2004).

A final clarification is necessary: the instruction to choose a collective position after the small group discussion is not intended to freeze the opinions that have been constructed, but rather to survey the different positions present in the class, in order to stimulate the debate. In no case are the students required to reach a consensus, and they are free to take the floor to justify a position other than the one voted by their group. They can change their mind, as their young opinion is under construction and will probably keep on evolving long after the two-hour activity.

- *The role of the facilitators*

Finally, let's focus on the role of the *YouTalk* facilitators. One of the first observations of researchers who were interested in the transposition of the scientific *café* to school was to point out the risks of triggering a question-and-answer dynamic between the expert and the students, which would take precedence over the establishment of an effective debate. One of the first choices of pedagogical design was therefore to remove the experts from the activity. The *Junior Café* was then led by a facilitator who did not have any expertise on the subject in the eyes of the students, in order to avoid the problem of *"the subjection of the students' discourse to the institutional relationship with the teacher, who is also often the manager of the debates"* (Albe, 2009b, p. 83). In fact, school routines, and in particular the classic dynamic of teacher-student exchanges on the Interrogate-Respond-Evaluate mode (Mehan, 1979) is contrary to the our pedagogical goals here. A change in the normative organization of the class is necessary to avoid this effect, which tends to limit students' engagement in argumentation, because " *what would be the point in trying to convince your classmates that your ideas has merit if the teacher would step in and solve the controversy with a simple yes or no?*" (Cornelius & Herrenkohl, 2004, p. 485). A further step was taken in the *YouTalk*, replacing the adult facilitators with a pair of high school student volunteers who received a day of training specifically for this purpose. During the training, two distinct functions were introduced: the main facilitator and the observer facilitator. The person in charge of the main moderation introduces the session and explains the rules of the activity, introduces each thematic phase, reads the questions, explains the "information desks" if necessary, goes to the different tables during the small group discussions, facilitates the whole-class debates, comments on the results of the opinion polls if necessary, and closes the *YouTalk*. The student observer facilitator introduces him/herself, monitors the length of each phase as the session goes, may give additional explanations on the "information desks", supports the small group discussions, observes who is speaking in the class and makes sure that each student has the opportunity to talk, and takes notes on the blackboard during the class debates on the OQ. Before the MQ is brought up again at the end of the *YouTalk*, these notes serve as a basis for a few minutes of synthesis aiming at reminding everyone of what has already been said and enriching the final socioscientific exploration.

1.3 Implementation of the YouTalk in 3 countries

1.3.1 Adaptations to the local context

The implementation of YouTalk in 3 different countries in 2011-2012 required slight adaptations to the canonical sequence described above. First of all, the concern to have several KQ directly related to students' daily lives implied to adapt them to the local context, depending on the available information, even if, for the purposes of comparison and consistency of the pedagogical settings, the MQ and OQ remained identical. In Mexico, the fieldwork was facilitated by a collaboration with the local non-profit organization of informal science education, *Pandillas Científicas*[10]. We owe the choice of the theme of drinking water to the fact that the implementation started in Mexico, where drinking water is a great issue. In fact, although it is futile to attempt an exhaustive enumeration of the differences between the distinct contexts investigated, we can nevertheless mention the fact that they are not *geographically* concerned in the same way by the matter of drinking wat. For example, Kenosha, the American city where we implemented the *YouTalk*, is located on the shores of Lake Michigan, one of the largest freshwater reservoirs on the planet.

The main adaptation of the activity to the local Mexican context was the individual voting modality. For technical reasons, it was not possible to set up an electronic vote, which was replaced by a paper questionnaire, not allowing the display in real time of the chosen response options, and therefore its use to stimulate the debate, but still making it possible to collect opinions for research purposes. This voting method also affected the principle of anonymity, as students tended to look at what their neighbor checked off.

For each *YouTalk* held in Mexico, the video corpus consists of 6 tracks:
- a general view of the class, associated with the sound of the microphone for the main facilitator;
- a recording of the projected screen with the computer's sound;
- 4 views each focused on a group of 3 to 5 students, coupled with the sound of a wired microphone placed in the middle of the table (out of a total of 4 to 6 groups in each session).

The specificity in the American school was that we could not implement the *YouTalk* in the initially planned time because the school required that we adapt to the usual duration of the class periods. As a result, the sequence was adapted to a duration of 88 minutes, with the removal of one QC from each thematic phase. In order to best respect the spirit of the YouTalk, we prioritized the KQ the most useful to apprehend the following OQ.

For the YouTalk held in Kenosha and Lyon, 4 video tracks are available:
- a general view of the class, coupled with the sound of the microphone of the main facilitator;
- a recording of the projected screen with the computer's sound;
- two views each focused on a group of 3 to 5 students, coupled with the sounds of wireless microphones worn by each student in the group (out of a total of 6 to 7 groups in each session).

---

[10] Literally "scientific gangs", facebook page of the organization: https://www.facebook.com/PandillasCientificasDeMexico/.

I would like to point out that the recording device for the table groups was highly visible, and that the students who sat at the table chose to be subject to this in-depth study. In total, 17 sessions were implemented: 5 in Contepec, 4 in Tehuacán, 4 in Kenosha, 4 in Lyon. Excluding the sessions where technical recording problems, and favoring those with the richest debates, ten of them constituted my thesis corpus, on which the analyses described here were conducted. I also sought to balance the weight of the different schools, finally retaining 2 sessions recorded in Contepec, 2 in Tehuacán, 3 in Kenosha, and 3 in Lyon. To view, organize, transcribe, and code this corpus, I used the *Transana* software. In addition to the video, the results of the opinion polls were also kept. When the vote was electronic, the *Activinspire* software automatically built up very precise digital traces. For example, it is possible to see how long it took each student to answer each question. In Mexico, the paper questionnaires were simply processed in a spreadsheet.

One final piece of context: let's quickly address the socioeconomic profile of the students whom we worked with.

1.3.2 Student Profiles

The student participants were between the ages of 12 and 15, and the facilitators were aged 16 to 19.

- *High school students leading the YouTalk*

I myself trained the high school students who volunteered to be facilitators for each of these four schools. Although the only explicit criterion that we defined for the choice of the facilitators was voluntariness, it was generally high achieving students who played this role. In the first Mexican city where we worked, Contepec, their school was independent of the participating students' school, and both schools were public. However, due to the size of the town, located in a rural area of Michoacán, they knew the participants well enough to call them by their first names. Only four high school students, one of whom was a girl, participated fully in the project, and four others came more out of curiosity at the beginning of the training, without showing again after lunch. For all the schools, the investment in the training, and then in the facilitation of one or two *YouTalk* sessions, implied being comfortable enough with the usual workload to manage this additional mission. In the second school visited in Mexico, in Tehuacán, Puebla, there were too many volunteers, and the teachers selected 6 boys and 4 girls, corresponding to the people that they thought as best suited for the task. In this small, higher-class, urban, private school, participation in the facilitation was highly supervised and the students could not give up the project once they started the training. The *YouTalk* that these students facilitated was for younger students in the same school. Again, they could call each other by their first name. It is also worth mentioning that in Contepec, the *YouYalk* was introduced as a very innovative educational practices as the school routines rather relied on a traditional top-down authoritative teaching model. In contrast, the private school visited in Tehucán was known for its dialogical tteaching methods, a context in which the *YouTalk* fit quite naturally.

In Kenosha (United States, Michigan), given the proximity of important academic deadlines, the teachers favored students with very good academic results, gathering a team of 6 facilitators. The intervention took place in a public, socially mixt school, recruiting mostly middle-class students. The student leaders then facilitated the *YouTalk* for younger students in the same school.

In Lyon, I was able to get in touch with the high school students through a physics teacher, and as a result, only students in the final year of the scientific curriculum participated in the training. About 20 students attended a presentation of the project during a lunch break, but in the end, only 7 (6 girls and 1 boy) attended the training and facilitated one or two session(s) of *YouTalk*. These figures can probably be explained by the specificity of the training methods offered in Lyon. Indeed, the school management only allowed the students to miss classes during the *cafés*, but not for the training, which took place on two Wednesday afternoons (a time when the students are usually not at school). The high school was next to the middle school where the YouTalk took place, both of which are located in the center of the city. However, due to the size of the schools, the students facilitating the YouTalk did not know the participants personally, neither in Kenosha nor in Lyon.

Table 2. Profile of students facilitating *YouTalk* sessions at the four schools.

City (Country)	Contepec (Mexico)	Tehuacán (Mexico)	Kenosha (USA)	Lyon (France)
Institution	Other, public	Same private school	Same public school	Other, public
Gender	3 boys, 1 girl	6 boys, 4 girls	6 girls	1 boy, 6 girls
Educational profile	Medium	Fair to good	Very good academic level	Scientific curriculum

- *Middle school students participating in the YouTalk sessions*

We collected some personal data on the social background of the YouTalk participants, based on their self-reports, with no surprises in relation to the geographical and institutional context. In Contepec, most of the students' fathers were artisans, merchants, or farmers, and most of the mothers were housewives or merchants. In Tehuacán, the occupations of the parents reported by the students were much more varied, and reflected the upper end of the social spectrum, including many executives, with the classic gendered predominance of mothers in the fields of care and education.

In Kenosha, more parents were reported as not working, both fathers and mothers. Employed parents are mostly in service sectors, at various levels. Finally, in Lyon, we find a similar pattern to that of Kenosha, but with more high-level positions and more parents working in education or research.

Generally speaking, the usual class groups of 25 to 30 students were used to organize the sessions, with the exception of Contepec, where we worked with half classes in order to keep the number of students at a reasonable level (about 20). The classrooms were set up with tables of 4 places, sometimes adapted for groups of 3 or 5 students, who chose themselves where to sit. This freedom led, unplanned, to gender non-mixing in one of the YouTalk sessions studied, held in Contepec, where an impromptu half-class group was made up exclusively of girls, while the main facilitator was a boy.

Table 3. Profile of students participating in *YouTalk Cafés* at the four schools.

City (Country)	Contepec (Mexico)	Tehuacán (Mexico)	Kenosha (USA)	Lyon (France)
School	Public, rural	Private, urban	Public, urban	Public, downtown
Main occupations of mothers	housewives, shopkeepers	higher or liberal professions, care and education fields	unemployed women, service sectors, various qualification levels	service sectors in high positions, education, research
Main occupations of fathers	craftsmen, shopkeepers, farmers	higher or liberal professions	unemployed, service sectors, various skill levels	service sector in high positions, education, research
Number of cafés studied	2	2	3	3

1.4 Learning from such a corpus: linguistics applied to educational practices

I carried out a detailed study of the students' arguments about the main question and the opinion questions in this large audiovisual corpus. As a linguist, my first concern was to give an account of the communicative acts of the students, in the language that they used, in order to follow and respect the movements of thought that they embodied. It is thus according to what emerged from their speech that I used specific concepts and methods coming from several disciplines in order to cast light on their reasoning under construction. As an educational researcher, this interest in their exchanges is based on a socioconstructivist conception of learning: it is through interaction that practices and representations can evolve. Such work can only be deeply empirical, strongly rooted in in-depth data analysis. This is why several of the chapters in this book offer detailed case studies based on excerpts from students' dialogues. Depending on the research questions, I adopt a varying granularity: I can discuss both the impact of a word or gesture in small-group argumentative interaction and what distinguishes the way the problem is viewed across different national corpora. Following Plantin (2018), I consider argumentative any situation in which students are asked to discuss a common question that raises doubt or disagreement (uncertainty about the answer, whether each person has no *a priori* opinion or whether several theses are in competition). However, this work is not limited to a collection of "cases". It also aims at identifying regularities, in order to better understand how the didactic context influences the argumentative practices, in a way that is more or less favourable to the development of thought. Indeed, the interest in making the results of this study accessible lies in the discovery of phenomena that are not limited to the framework of this *YouTalk*. The general contribution of this work is twofold. On the one hand, it provides conceptual and methodological tools for linguistic analysis that can be used to analyze other situations, and on the other hand, it informs the design of new pedagogical sequences for collaborative argumentation-based learning.

1.4.1 Scientific scope: studying socio-affective argumentation

A first originality of my approach consists in apprehending jointly the cognitive, social and emotional dimensions of argumentative interactions. Chapter 2 thus presents the descriptive, prescriptive and affective logics with which one argues, often intertwining them (2.1). Among the argumentative materials used to reason about SSC, knowledge deserves particular attention (2.2). Constitutive of the descriptive logic, the descriptive content used in these debates are very heterogeneous, based on distinct forms of legitimacy, which is why I speak of "knowledge-beliefs". I propose a grid allowing to characterize them according to five parameters: source, degree of generality, logical level, comparability to reference knowledge, didactic position with respect to reference knowledge. This tool allows us to better understand the use of knowledge about SSC. I also present in this first chapter an inventory of the prescriptive materials used by the students to argue in this corpus, distinguishing between norms of behavior or procedures and values (2.3). Beyond pointing out the importance and specificity of this prescriptive logic, it is also an opportunity to propose the main lines of an interactional grammar of argumentation on principles. What happens when we try to reason "in general"? More precisely, how does the fact of invoking a principle call for another one, through a set of interactional constraints on the possible refutation strategies? Emotions also appear as cognitive resources participating in the framing of the debate and in the argumentative orientation of the discourse (2.4). A case study shows how the emotional parameters[11] on the axes of intensity and valency can be simply applied to an extract of dialogue in order to identify the emotional position consubstantial with the argumentative conclusion defended.

Chapter 3 deals specifically with the questions of the conditions and manifestations of high-quality argumentation in small groups. For example, it is not enough to succeed in creating a real socio-cognitive conflict, which is not an easy task, but we must also ensure that it is resolved at the cognitive level, and not simply at the social level. In the extreme, this means that killing whoever disagrees with me or switching to another topic are effective strategies for resolving the relational conflict, but ineffective for exploring the cognitive problem at stake. Taking into account the role of emotions in this type of discussion, which have both social and cognitive functions, offers innovative avenues for reflection (3.4).

The second specificity of my enterprise is the respect of the debated objects, a concern that proved to be very fruitful for the creation of methodological tools of comparative argumentation, presented in chapter 4. Indeed, one does not discuss drinking water in the same way as the fashion of the handspinner. It is indeed the object of the debate that provides the keys to the argumentative scripts developed about it. When I tried to compare the arguments developed in my

---

[11]The adjective derived from the present participle is used here to account for the fact that these parameters are not in themselves impregnated with affect, but that they can be used, in the discourse, to confer on an object or a situation a certain emotional tone, aiming to arouse certain emotions about them.

different research fields, on the substantive level and not on the level of rhetorical styles, I had to equip myself with tools allowing me to define the contours of this object. Four clusters of clues, using three different methods, allowed me to highlight the predominance of argumentative scripts in each national corpus. The use of a coding schemes validated by robust inter-coder reliability scores, was efficient to identify the main thematic orientations of the debates and the metaphors used as cognitive models by the students to apprehend drinking water (4.1). A textometric approach to the discourse object "water" revealed the few recurrent viewpoints brought to it during the debates and their respective weight in the three national corpora (4.2.1). Coupled with voting data showing the competing response alternatives, these results converge to describe the typical argumentative scripts of each country, allowing for a comparison of the substance of the arguments (4.2).

1.4.2 Practical scope: fertilizing the debate

In terms of practical recommendations, the characterization of the cognitive resources used to argue about a SSC (chapter 2) and the centrality of the object of the debate (chapter 4) highlight the importance of the choice of the discussion theme and its formulation in order to develop a high-quality cognitive exploration. The interactional mechanisms of the argumentative use of principles (2.3), already show how a discussion is likely to get bogged down when it drifts towards the mere repetition of classical opposed scripts about a controversy. But it is essentially in chapter 3 that I address the pedagogical aspects of constructively debating about a controversial issue.

Indeed, students' argumentative practices are part of group work dynamics, which should be taken into account, as they structure their activity as much as the cognitive resources available to them. However, the engagement of a group of students into a constructive discussion dynamic allowing them to make the best use of these cognitive resources is not easy. I therefore begin by proposing a grid for evaluating the quality of group argumentation, considered as engagement into a genuine process of cognitive *exploration* of the question. Its application is illustrated by a few case studies (3.1). By comparing successful or emblematic cases of a type of discourse (*exploratory, cumulative* or *disputational*), I refine the definition of such an exploratory approach, and I highlight the factors likely to favour it or, on the contrary, to obstruct it. The emotions at stake in group argumentation, by their double role, social when it comes to the need for recognition and the preservation of faces, and cognitive, when affectively framing the debate, seem to be particularly important for the development of high-quality reasoning (3.2). On a practical level, controlling the emotional tone of the debate at the heart of instructional design, and seeking to make the social rules of the activity explicit throughout the process are likely to encourage a constructive exploration of the issue. Understanding and controlling the effects of personal challenges in such discussions (3.3) can also help to avoid or limit the sequences of *dispute* so that the argumentation remains constructive.

II.   Which cognitive resources to argue about SSC?

When aiming at exploring SSC, one cannot apprehend argumentation as a formal practice of presenting and organizing ideas according to a framework reduced to oratorical art. Argumentative practices about SSC actually develop from cognitive contents specific to the issue under discussion. According to the way in which they are elaborated and reconstructed in the debate, they can be based on three logics: the descriptive, the prescriptive, or the affective logic.

2.1   Reasoning: descriptive, prescriptive and affective logics

2.1.1 Argumentation as a multidimensional activity

Wrongly, an opposition has been maintained between, on the one hand, "rhetoric" (1), which has been devalued and reduced to a set of stylistic oratorical procedures that would base support for a claim by recourse to emotional instincts and relational pressures; and, on the other hand, "argumentation" (2), which is supposed to convince by appealing to human reason, a faculty assumed as independent from and morally higher than such instincts and pressures. Following Amossy (2006) and Plantin (2018), I defend a view of argumentation that goes beyond this opposition. While argumentation is certainly a cognitive activity, it is embodied and realized in linguistic, social, and, in the present case, interactional practices. Thus, the interest shown in this first chapter in the cognitive contents of argumentation does not aim to rehabilitate "the substance of debates" against something that would only constitute "the form" of it, the two being, in authentic discourses, inseparably intertwined. In order to argue, students carry out discursive 'work' in order to construct and defend their position, woven by simultaneous recourse to classical argumentative reasoning and to the argumentativity of language, in the sense of Anscombre and Ducrot (Anscombre and Ducrot, 1997 [1983]). By constructing and then elaborating or transforming their discourse objects, students produce schematizations (Grize, 1990; Grize, 1996) that form the basis of their arguments. Actually, their reasoning, exploiting either causality, definition/categorization, or analogy is to be found both in the very properties of the linguistic material, and in the relationships between the things to which such material refers. By studying the cognitive contents used by the students, I rather seek to point out three different logics of reasoning that are at work in these discourses. In my dissertation, I already proposed a schema of students' argumentative work showing that it relies on three types of cognitive resources: knowledge, norms and emotions. In his impressive work on values in argumentation, based on a huge corpus of polemical discourses on social issues, Guerrini also distinguished three poles of fundamental discursive resources used to argue: emotions, stabilized knowledge, factual data & beliefs, values (Guerrini, 2015, p. 471). Today, it seems more appropriate to me to speak of descriptive, prescriptive and affective logics as constituting the three pillars of argumentative reasoning about SSC. These processes are cognitive insofar as they are logico-discursive, both genuine thought movements and linguistic constructs, a property of ordinary reasoning well described by Grize:

> Natural logic can be defined as the study of the logico-discursive operations that allow the construction and reconstruction of a schematization. The double adjective underlines the fact that we are in the presence of operations of thought, but only insofar as these are expressed through discursive activities. (Grize, 1996, p. 65).

The descriptive logic is based on the use of elements presented as facts, whose veracity is not likely to be questioned in the debate. Participants *a priori* share the belief that these elements are sufficiently adequate with reality to provide a satisfactory description of it, as a common basis for discussion. Of course, the descriptive logic used in an argumentative situation cannot be "neutral": among all the facts that could be mentioned in the landscape of a problem, only some are made explicit, at a given moment in the discussion, when they appear appropriate to defend one's view.

The prescriptive logic relies on the use of norms or values to support or challenge the arguments that emerge in the debate by depicting them as leading to desirable or undesirable situations or behaviors. This evaluation can take place at the discursive or metadiscursive level. At the metadiscursive level, ideas are assessed referring to considerations on what a good argument is, in general, and/or in the particular context of such debate. The study of this type of metadiscursive assessment is close to Doury's research on ordinary argumentative norms (2004), and goes beyond the consideration of expected communicative behavior, also including epistemic concerns. Of course, the norms then used as reference principles may either be accepted by all the debaters, or themselves become objects of discussion.

Affective logic consists in arguing by associating a positive emotional tone to the point of view defended and/or by associating a negative emotional tone to the opposing point(s) of view. When debating on social issues, no discourse is emotionally neutral, even if pretending to speak without affect may be an argumentative strategy in some contexts. Two things constrain the argumentative use of affective logic. First, the discussed issue, in itself, the way it is introduced for discussion, already provides a specific emotional tone. This is rather high in the case of drinking water management, since it is a vital resource. Moreover, and this is quite striking when we have a corpus of numerous debates on the same issue, during a given debate, a basic emotional color is established that "sets the tone" and delimits the field of possibilities in terms of affective argumentation. For instance, if all the participants feel concerned by the problem and frame it as very important and urgent, a discursive strategy of "cooling" the atmosphere by disengagement or presenting it as a minor immediate concerns will be difficult, if not impossible, to develop.

2.1.2 Distinguishing between description, prescription and affect

When we consider argumentation as aiming to have others think, say or act specifically on a given question, all the cognitive elements on which prior agreement exist are great resources, because they constitute a common basis from which to get others accept one's conclusion. Perelman and Olbrechts-Tyteca have classified these "objects of agreement" into two categories: those relating to the *real* (facts, truths, presumptions), and those relating to the *preferable* (values, hierarchies, "places of the preferable") (Perelman and Olbrechts-Tyteca, 1988 [1958], p. 88). Emotions, almost absent from their work, inspite of their now widely-recognized cognitive dimension (Damasio, 1995), should also be included among the resources on which we rely to assess the legitimacy of a proposition. Thus, depending how we describe reality, how we feel about it, and how we consider it as more or less acceptable, we produce a discourse with a specific argumentative orientation that claims for a preferred answer to the question at stake. I use the term descriptive logic when the "objects of agreement" on which this argumentative reasoning is based are elements of knowledge-belief; of prescriptive logic when they are values or norms of behaviour; and of affective logic when they are emotional parameters. Let's now define more precisely each of these elementary ressources of argumentation.

- *Knowledge & beliefs*
Taking into account beliefs together with knowledge in descriptive logic is essential to understand ordinary, authentic arguments. Indeed, in argumentative discourse, knowledge and beliefs about reality function in a similar way, as *objects of agreement* allowing to assess the likeability of propositions. As such, they are presented as

unquestionable facts. Such "status of fact" confers them the force to support a thesis as an attested element, exempt from any contestation: "The fact as premise is an uncontroversial fact" (Perelman & Olbrechts-Tyteca, 1988, p. 91).

But, of course, what constitutes an agreement, between the participants, at a given moment, about the world, may be more or less realistic regarding to the canons of the legitimate construction of knowledge. These elements can thus be considered as belonging to a *continuum* ranging from "knowledge" in the strict sense (*episteme*) to beliefs that were not validated by any scientific control (*doxa*). This broad category thus includes ideas about reality of varying epistemic status. Useful for the analysis of argumentative interaction, it is limited from a scientific point of view, when, in order to construct rigorous knowledge, it is important to discriminate between diverging representations of the world according to their mode of elaboration. From a didactic perspective, it is also essential to clearly distinguish scientific theories from "beliefs" that have not been rigorously investigated (cf. 1.1.2).

- *Norms & Values*

Prescriptive logic consists in arguing from *objects of agreement* about what is desirable (values) and what is socially correct (norms of behavior, generally based on such values). It allows one to assess the acceptability of a proposition with reference to the social norms and commonly recognized values.

Plantin distinguishes the norm as an element of description (the statisticians' norm), which refers rather to an element of knowledge-belief; and the norm as a prescription. It is in this second sense that we use this term, as "the injunction of an obligation, which is expressed by a rule whose content is the responsibility of the institution or particular domain concerned" (own translation, 2018, p. 410): morality, linguistic rules, rational behavior, traffic regulations, etc. When someone, during a debate, refers to an argumentative norm, he or she shifts to a metadiscourse to assess statements according to the behavior(s) of the person(s) who are stating them. These argumentative norms can be spontaneous or they can result from a technique aiming at regulating exchanges in order to reach a rational agreement (goal of the pragma-dialectics).

Values orientates the argumentative discourse by giving it a goal. They allow us to formulate value judgments about an object, a person, a decision, by evaluating their quality (good, beautiful, fair, etc.). The reference to values in the discourse can be explicit, or remain implicit. The value-words (freedom, solidarity, beauty) are the most explicit form of the use of values to argue. They can be invoked ("in the name of freedom"), called upon, advocated, argued. The values can also be appealed to via adjectives (*free*), adverbs (*freely*), verbs (*liberate*). More implicitly, all the lexical material can be used axiologically. It is possible, with a few *a posteriori* control procedures (is x considered a value? can we say "the value of x"?), to make such axiological perspective visible by asking the question "in the name of what?" is a given opinion defended.

Values are often in competition and cannot always all be satisfied at the same time. Priorities must then be established, incompatibilities overcome, by making compromises. Of course, at some point in the debate, certain values may no longer be objects of agreement. Episodes of disagreement about what is desirable may uncover deep divergences, revealing what really matters to each sides (Guerrini, 2015), or may lead to discussions about how to prioritize between shared values.

- *Affects & emotions*

Emotions have recently been reconsidered as legitimate argumentative resources (Plantin, 2011, Polo et al., 2013b), after long being considered sources of (potentially) fallacious reasoning (e.g., Hamblin, 1970; Walton 1992). In cognitive science and education research, the 'affective turn' overcame the traditional emotion/reason dichotomy (e.g., Damasio 1995; Baker, Andriessen, Järvelä, 2013). However, authors interested in emotions in argumentation study diverse objects, a panorama that relies on the following pairs: experienced or expressed emotions; long-term emotional construct (e.g., trust in a working team) or emotional variations of shorter duration (e.g., tension/relaxation); individual or collective emotions (Polo et al. , 2016b, Polo et al. , 2017a).

In psychology authors traditionally distinguish emotions from feelings and moods, on the one hand, and affects, on the other hand (Cosnier, 1994). Feelings and moods would last longer, and rely on intellectual perception or even construction. On the contrary, emotions would be brief and more "spontaneous", associated with a stimulus situation or event. The distinction between affects and emotions is based on the appraisal of the event and the associated feeling. The term *affect* tends to be used for undetermined feelings, and *emotion* tends to be used for more specific, labelled ones (*joy, disappointment*, etc.) (Plantin, 2015). Some emotions are typical to the argumentative situation. In education, part of the literature focuses on emotions associated with the socio-cognitive conflict: tension raised by disagreement, fear of losing face, relaxation strategies, and a climate fostering heuristic argumentation (e.g., Andriessen Pardijs, Baker, 2013; Sins & Karlgren, 2013; Allwood, Traum, Jokinen, 2000). In the field of rhetorics, some strategies have been considered, since Aristotle (*Rhetoric*, Book II, Chapters 1-10), to be closely related to the use of "passions." All these objects (moods, feelings, affects, emotions, passions) are included in the category "affects and emotions" that I consider to study affective logic.

Indeed, when it comes to the argumentative use of emotional feelings, it does not matter whether we represent them as long or short, with a rather intellectual or corporal focus, whether we name them precisely or display a vague affect. What counts above all is to understand how such resource constitutes an affective "object of agreement" that supports the defense of an argumentative claim.

2.1.3 Relations between these elementary argumentative resources

It is essential to specify that, in ordinary arguments, descriptive, prescriptive and affective logics are closely intertwined. The same statement, the same speech turn can jointly use elements of knowledge and beliefs, norms and values, and affects and emotions. This is why I propose to represent them, in a global visualization of the elementary argumentative resources, as linked (cf. figure 3). Study their relations could, by itself, constitute a whole book. I limit myself here to a few remarks based on the definitions of these three categories that I gave above.

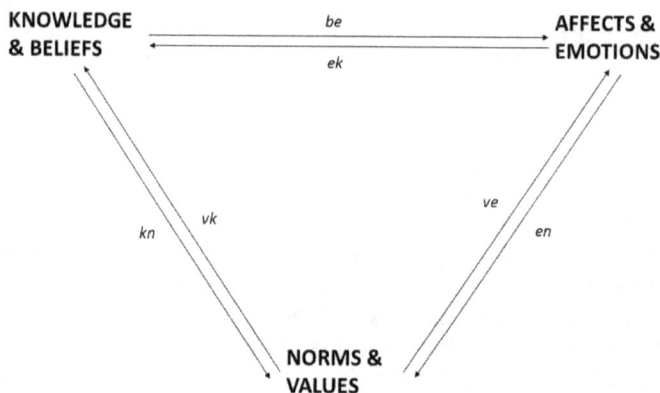

Figure 3. Examples of relationships between descriptive, prescriptive, and affective argumentative resources.

An element treated in the discourse as a fact can be used to support a norm (relation noted *kn* of a knowledge *k* towards a norm *n*). For example, its hygienic virtue can be mentioned to defend circumcision. An incident reported in the media will confirm the norm that one should not speaking to strangers. Conversely, a value may appear in a discourse through the axiological judgement made on an event (relation noted *vs* from a value *v* to a knowledge *k*). We will find it *deplorable* to discover that Freud committed scientific fraud, because we share values as the search for truth, honesty, and epistemological rigor. Finding an alternative that conforms to a shared value will be pleasant (relation noted *ve* from a value *v* to an emotion *e*). Thus, merit can be appealed to in order to legitimize social inequalities, giving them a softer emotional tone than if they were appraised by the value of social justice. Sometimes, the use of the lexicon of emotion functions as an appeal to norms, as for this job applicant who is told that it is *disappointing* that he did not wear a tie (relation noted *en* from an emotion *e* to a norm *n*). Finally, the *emotional schematization* exploits knowledge & beliefs. Thus, talking about the fate of "our European neighbors in Ukrania", a more emotionally intense expression than "Ukranian nationals", for example, introduces the fact that Ukrania is located in Europe (relation noted *ek* from an emotion *e* to a knowledge *k*). Similarly, the quasi–religious belief in a global human community may lead to the designation of refugees as "human brothers" rather than "candidates for immigration" (a relationship noted *be* a belief to an emotion).

Arguing relies on descriptive, prescriptive and affective logics, often intertwined in ordinary discourse. It can be useful for research analysis and didactical practice to distinguish between the resources used for each of these logics (knowledge-beliefs, norms and values, affects and emotions), partly because they differ in terms of epistemic status. However, none of the three types of argumentative resources is *a priori* cognitively higher or lower than the others, and any of them may play a great role to successfully explore a SSC.

2.2 The descriptive: what place for knowledge?[1]

2.2.1 Knowing-believing continuum in ordinary conversations

One of the teachers' reticence regarding lesson on SSC is the fear that putting debating about them would only lead to sterile battles of opinion (Albe, 2009a). Actually, tools to characterize and evaluate the content knowledge built in such settings need to be developped. This is all the more delicate since the objects addressed and the pedagogical setting itself are intended to mobilize students' knowledge and beliefs of diverse epistemic status and coming from diverging sources. Indeed, to bring SSC into school is to invite profane knowledge to the classroom. Thus, it would be inappropriate to try to characterize these hybrid descroiptive elements only referring to the traditional category of "knowledge", defined in a restrictive way. In reality, the objects that interest us here are rather "units of knowledge-belief". These elements do not correspond to "learnt knowledge", in the sense of institutionalized knowledge validated by the scientific community. Nor is it "school knowledge", resulting from a didactic transposition (Chevallard, 1992) of scientific facts. If the students, in these *cafés*, do rely on information that they present as factual truths, these belong rather to what Beitone and Legardez (1995) call the "system of representations-knowledge" of the actors, described by Simonneaux in these terms: "it is an aggregate comprising opinions, beliefs, attitudes, information from various sources (including popular science), parts of previous school learning, social representations" (2006, p. 46).

We are therefore interested here in these elements of description of the world, presented as facts in students' discourse, by making the hypothesis of a continuum between knowledge and beliefs, and without *a priori* defining their epistemic status. Indeed, Vignaux (1976) has rightly emphasized the constructed nature of "facts", and *science studies* (e.g. Latour and Woolgar, 1979) have also described the "socially constructed" character of scientific activity, an institution producing "facts" by excellence. But it is not because they result from a social "construction" that facts cannot play the role of warrants of

---

[1]An extended version of this section was published in French in the following article Polo, C., C. Plantin, K. Lund, G. P. Niccolai (2016), Savoirs mobilisés par les élèves dans des cafés science : grille de caractérisation issue d'une étude internationale, Recherches en Didactique des Sciences et des Technologies 13, 193–220 (https://halshs.archives-ouvertes.fr/halshs-01503177).

arguments. What is decisive here for considering an element of discourse as descriptive is thus the possibility of making it appear, in support of a thesis, as an accepted "fact" that is not to be demonstrated.

So how can we put order into the abundance of very heterogeneous elements of description that emerge in these *cafés*? The challenge is to identify structuring dimensions that will allow us to better understand how they function in students' argumentative discourse. What conceptual distinctions can be useful for such an undertaking, both from the perspective of identifying students' reasoning strategies and comparing them to reference knowledge? From a didactic perspective, it is indeed important that theories attested by communities of scholars be distinguished from "beliefs" that have not been the object of rigorous investigation. Rather than trying to artificially "fit" the students' statements into pre-existing theoretical categories, I have chosen to empirically identify recurrent elements. First of all, the knowledge-beliefs were grouped by content, making it possible to identify the recurrence of similar information, and to draw up an inventory. This first confrontation with the data led, not surprisingly, to the observation of a very great heterogeneity among the elements listed, at different levels: size (duration, verbal material), level and genre of language, degree of explicitness (specialized vocabulary or paraphrase), etc. I have thus gradually defined five analytical categories.

2.2.2 Apprehending heterogeneous elements of knowing-believing: 5 characteristics

I have developed a classification of the elements of knowing-believing mobilized in these debates according to 5 dimensions: logical level, degree of generality, comparison to reference knowledge, didactical position towards reference knowledge, sources of information. Table 4 summarizes these different dimensions by including a few examples taken from the corpus.

Table 4. Grid characterizing the elements of knowing–believing used to discuss a SSC.

Source(s)	Logical level	Degree of generality	Relation to reference knowledge	
			Comparability	Didactic position
1. Direct experience	Objects definition of beings and objects	Precise	A-doxal	Quasi-referential
		a bottle of water at the store at the corner costs 1 $	*there's a hole in the ozone layer*	*the rarer the water is, the most expensive it becomes*
	Earth is mostly recovered by water			
2. Testimony				Misleading
3. School learning				*water is gonna be the new currency*
4. Previous KQ				Erroneous
				Dubai is in India
5. Relatives				
6. Media	Relations between things	General	Doxal	
		what is true today will be true tomorrow	*people are selfish*	
	we need water to live			

- *Logical level*

A first component, allowing us to classify these very heterogeneous elements, has to do with the logical level considered. In classical logic, argumentation is defined by three "operations of the mind", working at different levels: apprehension concerns the definition of beings and objects; judgment affirms something about them, leading to a proposition; and reasoning links propositions together. These operations correspond to the linguistic processes of reference, predication, and argumentation (Plantin, 2018, p. 282). To explain the construction of schematizations, which are intrinsically argumentative, Grize also breaks them down into successive operations:

1. operation of constitution, that shape primitive notions into discourse objects or classes of objects;

2. operation of determination, leading to construction of a predicate; operation of enunciation (a subject uses a predicate in discourse), leading to the statements; operation of configuration establishing relations between statements (Plantin, *ibid.*, p. 424; Grize, 1990, pp. 65-77, own translation).

The elements identified in our corpus belong to various "logical levels" of cognitivo–linguistic operations. Thus, we can distinguish descriptive elements concerning objects from those concerning relations. The

40

logical level of 'objects' includes a definition of "beings" and "objects", which is not limited to reference, but can also consist of predicates, and even encompass potential relations with other beings or objects when they correspond to *topoi* constituting the identity of the "object" or "being" in question. These descriptive elements concern an object and/or what Grize calls its "bundle": "I call the bundle of an object a set of aspects normally attached to the object. Its elements are of three kinds: properties, relations, and action patterns." (Grize, 1990, p. 78). For example, the following statement constitutes a descriptive element relating to an object: "the sea is blue".

Other knowledge-beliefs correspond to another logical level, and establish links between these beings and/or these objects. The judgments about these objects and/or beings are part of it, in the sense that they relate these beings or objects to the enunciator. For example, the following statement proceeds from a higher logical level, because it is a descriptive element concerning a relation between beings and an object: "fish live in the sea".

- *Degree of generality*
The items listed also vary in their level of generality. Some are very specific, such as the price of a bottle of water at the local grocery store, while others deal with global issues, such as global trends in water prices.

- *Comparison with reference knowledge*
In the same way, the knowledge-beliefs differ by their characteristic of being more or less translatable into the language of reference knowledge, and thus more or less comparable to the scientific stabilized knowledge. Some of them cannot be easily compared to it: their vague nature makes it impossible to say whether they are compatible with this knowledge, without excessive interpretation. These elements are called *doxal*, in reference to the notion of *doxa*. They sometimes belong to non-scientific domains (metaphysics, theology, religion), where the "truths" are those of communities of believers or families of minds. The statement "people are selfish" is typically a doxal belief that cannot be scientifically assessed, unless it were considerably revised.

- *Didactic position towards reference knowledge*
Fourth structuring axis, when the descriptive elements lend themselves to a translation or approximation in the language of reference knowledge, they can constitute either a favourable ground or an obstacle to learn this knowledge. Of course, this dichotomy is relative and depends on the learning goals and didactic contexts. It is not a question of defining a descriptive element as being true or false

by nature, but as more or less facilitating the learning of the corresponding reference knowledge, when it could be identified.

As soon as a descriptive element can easily be compared with reference knowledge, and facilitates its adquisition, it is qualified as *quasi-referential*. For example, to say that human beings have 5 fingers on each hand is not always absolutely correct, but it is an acceptable approximation of things in the light of reference knowledge. Conversely, certain descriptive elements are likely to hinder the learning of this corresponding reference knowledge. When they constitute, in themselves, obvious untruths, they are qualified as *erroneous*. For example, to say that human beings have four fingers on each foot is wrong. However, most things that are not *quasi-referential* are not *erroneous*. Many of them, regardless of their degree of truthfulness, can constitute obstacles to the learning of reference knowledge, because they convey or reinforce simplistic representations that make it difficult to aquire a more complex representation of the concerned object. They are then qualified as *misleading*. This is the case of all Newtonian physics when considering other reference frames than the Earth. For example, to say that the ball "falls" on the ground can be *misleading* when one tries to teach that it is "attracted" by the Earth.

- *Sources of information*

Finally, the descriptive elements can come from a multiplicity of sources, either mentioned or inferred. Its source constitutes one of the rare indices of the positioning of an information on the knowledge-belief axis. Indeed, knowledge and beliefs draw their authority from their sources and their elaboration, as well as from the relationship that is established with the person who uses them. However, the source of the facts mentioned by the students is rarely made explicit. Sometimes, they prefer to claim them as if they were obvious. Such a strategy limits argumentation in a certain way, since such obvious statements do not need justifications, and exclude any possibility of discussion.

In our corpus, six sources were identified:
1. direct experience, in accordance with Albe's observation: "personal experience appears as a means of interpreting and problematizing the controversy, a reference for arguing and justifying a position" (2006, p. 105);
2. the testimony of others[2], which ensures another form of proximity with the object, because, by exposing his experience ("I saw it", "I was there"), the witness brings a direct proof of the facts

---

[2]Testimony is a spectific form of appeal to authority (Plantin, 2018, p. 563).

3. "parts of previous school learning" (Simonneaux, 2006, p. 46);
4. parts of knowledge covered in previous KQs of the *café*;
5. parents and family;
6. the media[3].

The logical and generality levels inform us on the students' forms of reasoning, and in particular of their ability to shift from examples to abstraction and generalization. The other three dimensions concern the relation between the elements that they use as "facts" and reference knowledge, and make it possible to question the process of transitionning from one to the other.

2.2.3 Descriptive elements used by the students

We will first present the most frequent descriptive elements in the corpus in order to show how they can be identified in the students' speeches. The results of the inventory carried out by national corpus will then be put into perspective using the analytical grid introduced above.

• *Inventory based on authentic utterances*
In my dissertation, I conducted a comprehensive inventory of the elements of knowledge and belief used in the debates about the opinion questions and the main question (Polo, 2014, pp. 253–281). Whenever possible, I did not produce a paraphrase to refer to an item, but instead took up one of its occurrences in the form of an emblematic authentic statement uttered by a student. This access to the students' discourse itself illustrates the how such an analysis works, and the very heterogeneous nature of the data studied. In table 5, I reproduce the descriptive appearing at least 3 times in at least one of the three national sub-corpora, starting with the most frequent for each national corpus.

In the Mexican corpus, the most frequent descriptive element is "the rich will have more access to water". Five occurrences correspond to the public school, with for example[4]:

---

[3]If these last two sources can appear, in reference to the theory of argumentation, as two instances of a similar process of appeal to authority, it is useful to differentiate them here for didactic reasons. Indeed, it is not the same thing to take up a discourse directly heard at home from a parent's mouth as to draw elements from a discourse belonging to a mediated cultural production, whatever the medium (television, press, cinema, song, etc.).
[4]The following excerpts, like all those that will be quoted in this book, correspond to transcripts of the utterances made according to simplified conventions based on the standards developed by the

MYR   sólo los ricos van a tener (only the rich are going to have it)

For the private school, 19 statements refer to it, such as:

MIG   sólo la gente rica va a tener más disponibilidad al agua\ (only rich people will have more access to water)

This element is also the most frequent in the American corpus, with 21 occurrences, including:

BRE   the riches get the water\ and: no matter where you are on the globe if it's a poor area unfortunately

It is alsto found in the French data (6 occurrences):

PHI   les pauvres n'auront pas de auront pas d'eau potable `fin ou de l'eau de la mauvaise qualité et ceux qui [...] sont riches eh ben ils auront d'la bonne qualité\ (the poor will not have any drinking water well or water of bad quality and those who [...] are rich well they will have good quality\)

Table 5. Inventory of the elements of knowing–believing used in the YouTalk corpus.

Knowledge-beliefs appearing at least 3 times in a school	Mx	US	En	Total
The rich will have more access to water (bigger quantity or better quality)	24	21	6	51
Old habits die hard	4	19	17	40
We need water to live	14	13	5	32
We will have more or less water later depending on whether we save it today	22	4	3	29
Not all geographical areas have the same water resources	4	14	7	25
Water prices will increase	12	1	9	22
What is true today will be true tomorrow	3	7	12	22
Seawater is desalinable	3	13	7	21
Water resources are decreasing	20	0	1	21
Humanity adapts to nature	2	10	6	18
Desalination of sea water is expensive	1	10	5	16

ICOR group http://icar.univ-lyon2.fr/projets/corinte/bandeau_droit/convention_icor.htm). They are presented in the appendix.

	15	1	0	16
virtual water	15	1	0	16
What is rare is expensive	8	2	3	13
Some countries are richer than others	0	1	10	11
In Africa, there is not much water	2	4	4	10
In [own place] we have plenty of water	0	10	0	10
We pollute the water	4	3	2	9
Going to the bathroom uses a lot of water	5	0	4	9
People would abuse water if it were free	2	4	3	9
Nature is changing, degrading	1	4	3	8
The Earth is mostly covered by seas	0	6	2	8
Water is wasted	6	1	1	8
Washing consumes a lot of water	6	0	2	8
Not everyone will be able to keep up	1	3	3	7
Water will become the new currency	7	0	0	7
Making water safe to drink has an economic cost	4	2	1	7
Water can be recycled or purified	3	3	0	6
There are people who have no water	3	2	1	6
The definition of dry toilets	3	0	2	5
It's impossible to control how people use water	1	0	3	4
In Africa they have no money	0	0	3	3
There are wealth inequalities within countries	0	0	3	3

The second most frequent descriptive element, in the Mexican corpus, is "water resources are decreasing". Here is one of the 7 occurrences from the public school:

> MAR   en un futuro ya el agua (in the future the water will already)
>
> MAR   ((hand gestures downward))

There are also 13 occurrences in the private school, including:

> NAT   el agua ya se está acabando (the water is already running out)

Less present in the French corpus, it appears there under the following form:

> DEL   comme quoi on va plus avoir d'eau (that we would not have water anymore)

Finally, the concept of virtual water[5] is also very present, especially in the Mexican private school, for which 13 occurrences have been listed, including:

> RAU   lo vimos en la tabla anterior\en la tabla de: cuanto de litros de agua [...] por [...] un recurso en específico\ (we saw it in the previous table: how

---

[5]This relational descriptive element can be formulated as follows: more or less water is used to produce other things that we consume.

45

many liters of water [...] for [...] a specific resource)

This element also appears in the Mexican public school:

ARM   la carne es la que tiene más agua (the meat is the one using the most water)

It is found in the American school:

ERI   the production of meat and all it represents

The second most common element in the U.S. field is "old habits die hard" (19 occurrences). It appears for example as follows:

ETS   it would be very hard to get people [...] to actually stop you know and change our whole lifestyle

It is the most frequent descriptive element in the French corpus, with 17 occurrences, such as:

ERI   on était habitué à notre mode de vie donc y a personne qui va faire des efforts (we were used to our way of life so nobody will make any effort)

This element does not appear in the Mexican public school, but is present in the private school, for example:

MYR   en méxico la mayoría de lo que consumes es la carne es lo típico [...] no vamos a dejarlo (in mexico the majority of what you eat is meat it is typical [...] we are not going to stop)

The third most frequent element in the American corpus is "not all geographical areas have the same water resources" (14 occurrences), either expressed in quantitative or qualitative terms, as in the following examples:

JIM   a place that's more populated with water than other places
NAN   for some countries the water is like bad quality

It is found in the Mexican corpus, both in the public school:

PAU   en el norte tenemos agua [...] y en otras regiones no van a tener ni para vivir\ (in the north we have water [...] and in the other regions they won't have [water] even to live)

and in the private school:

> ALE   en otros países así ya falta el agua (in other countries like water is already lacking)

This element also appears in the French corpus:

> ANG   y'a des pays où y'a beaucoup moins d'eau que: dans d'autres (there are countries where there is much less water than: in others)

For the French corpus, the second most represented element is "what is true today will be true tomorrow" (12 occurrences), with statements such as:

> JER   parce que: à l'heure qu'il est c'est par rapport à A (because: at the present time it is in relation to A)

This element, to a lesser extent, is also mentioned in Mexico, in the public school:

> MYR   como va de avanzada ahorita la tecnología puedamos: puedan encontrar algo (as the technology goes advanced today that we can: that they can find something)

One occurrence was noted in the Mexican private school:

> GAS   el precio del agua va a empezar a incrementarse porque se va a empezar a agotar bueno se está agotando el agua y creo que creo que actualmente se está aumentando el precio del agua (the price of water is going to begin to increase because it is going to begin to exhaust well water is already exhausting and i believe that i believe that currently the price of water is increasing)

This element is also present in the American school:

> BRE   well because it's already like that

Finally, the third most frequent element in the French data is "some countries are richer than others" (10 occurrences), in forms such as:

> LEA   y'a des pays pauvres et y'a des pays riches\ (there are poor countries and there are rich countries\)

It does not appear in the Mexican corpus, and only once in the American corpus:

> MIC   the US since it's like a superpower

It is also worth mentioning that the most frequent elements tend to be used first in a specific question, or in a small group, before being taken up in the whole class discussion of the MQ. In addition, all the descriptive elements illustrated here are about relationships, between beings and/or objects: the rich – water; water resources – time; a product – water used to make it; habits – time; geographical areas, understood in terms of their water resources, compared to each other; veracity – time; countries, understood in terms of their wealth, to each other.

• *Applying the grid to characterize elements of knowing–believing*
In this section, the main characteristics of the exhaustive inventory carried out on each of the national corpora (including those present in less than 3 occurrences in one country) are listed, criterion by criterion. This analysis illustrates the contribution of the analytical grid of the elements of knowledge–belief to make sense of the abundance of heterogeneous descriptive elements stated by the students.

In the Mexican sub-corpus as a whole, we observe a predominance, in terms of logical levels, of elements dealing with relations over those describing objects. Among the elements dealing with objects, we find, for example, the fact that sea water can be made drinkable, the price of a bottle of water at the local grocery store, the definition of dry toilets, the existence of a diversity of cultures, and the mention of tricks to use less water. But most of the facts mentioned establish relationships between several beings or objects, for example the fact that one needs water to live, that more or less water is "incorporated" into everyday products, or that different factors have an influence on the price of water. This predominance of relations over objects also appears in the American corpus, even though there are 26 objects mentioned, for example "the Earth is mostly covered by seas"; "there are techniques for cleaning up water"; "there is a new pipeline project"; "water can change state [from liquid to solid for instance]". It is interesting to note that a good number of these objects (18/26) appears only once. A specificity of the French corpus lies sharing many facts about objects, which represent more than a third of the total of the descriptive elements there (37/94). They range, for example, from the mention of the Fukushima disaster to that of the Greek economic crisis, includes deforestation, the constitution of matter into molecules and atoms, the fact that tap water is not drinkable in Turkey, and the definitions of "consumer society", "dry toilets" and "groundwater". The other descriptive elements focus on relationships, for example "you can't find a buyer for something that is given for free elsewhere" or "if water were free, there would be abuse".

On the other hand, the most frequent descriptive elements are rather general, even if they can sometimes group together statements presenting a less general concrete information. In the two Mexican schools, the elements of knowledge–belief listed can be classified into three groups, according to their degree of generality. The first group corresponds to a very low level of generality, concerning precise examples of daily life, formulated in small groups. There are many examples of how to save water in everyday life. This contribution is emblematic:

> EDU y cuando te bañes en lo que se enfrie el agua poner una cuveta (and when you wash waiting for the water to cool down put a basin)

Here, the use of precise and concrete information seems to be a prerequisite for a potential subsequent generalization, or a support for implicitly conveying the principle in question. Indeed, in the whole class, the facts used tend to present a higher level of generality, the passage to the "public debate" constituting a form of institutionalization, implying a certain transposition of descriptive elements, to use the terms of didactics. We find, for example, in this second group:

> FAB en el futuro como va a haber escasez así que se va a subir el precio\ (in the future as there is going to be a shortage so the price is going to increase)

Finally, the third group of occurrences is rather hybrid in its level of generality. It consists of specific information, which appears in the whole class, but is presented from the outset as examples intended to explain or support a general reasoning.

For the American school, the fact that most of the elements appear only once is related to the low degree of generality of most of them, such as the one dealing with flushing toilets in Australia, or the one about the new pipeline project from Lake Michigan. In contrast, the most frequent descriptive items have a very general scope. The illustrative cases of the general law "not all geographical areas have the same water resources", concerning the situations of Africa and the students' city, are to be compared with the third group of occurrences highlighted in the Mexican field, namely the exemplification of a general reasoning.

The items listed in the French corpus vary greatly in their degree of generality. We can thus define a *continuum*, ranging from very precise information, such as the local price of water, to broad generalities, such as the element "there will always be rich people". In between, we find definitions of specific terms (dry toilets, groundwater, etc.). At the broad generalities pole are facts formulated in both group and whole class discussions.

The question of the relationship to reference knowledge is twofold. On the one hand, it is a matter of studying the comparability of the descriptive elements to this knowledge, their more or less *doxal* character. On the other hand, it consists in indicating, when they are comparable to it, whether they constitute a favourable ground or rather an obstacle to learning such reference knowledge. It goes without saying that this last element of characterization depends on the domain considered. Very few facts formulated in the Mexican public school are easily "translatable" into the language of reference knowledge. Thus, the very frequent elements "the rich will have more access to water" and "water resources are decreasing" are *doxal*. This last example shows a regularity concerning the evolution of water resources over time that cannot be understood in terms of reference knowledge without multiple transformations and interpretation. Indeed, a scientific evaluation of its veracity would require a reformulation with a series of hypotheses corresponding to an extensive interpretation of the students' statements, concerning the type of resource and water in question, the spatio-temporal framework of the evolution considered, etc. Depending on the assumptions made, this could be considered more or less correct or incorrect. These *doxal* elements represent 10/59 of the knowledge–beliefs listed for the second Mexican school, or 16.9%. In comparison, in the American school, 12/76 items, or 15.8%, are *doxal*. However, this is the case of the very frequent "habits die hard". The properly *doxal* elements are also a minority in the French corpus (18/94 or 19.1%). It is interesting to note that they often correspond to radical phrasing, characterized by the absence of mitigation, a linguistic form that tends to move the statements towards the "belief" side of the *continuum*.

> AGN y'a toujours des gens plus riches (there's always richer people)

Thus, to speak of "people" rather than "some people" is to state a belief about human nature rather than to relate a behaviour that may have been witnessed about some individuals. Similarly, the use of the adverb "always" seems to be associated with the formulation of a *doxal-type* element. The study of the very frequent element in the French corpus "what is true today will be true tomorrow" is delicate. On the one hand, it is generally a belief, of *doxal* nature. However, it can correspond, in certain precise didactic contexts, to an epistemological principle, such as actualism in geology, and then be classified as *non-doxal* and *quasi-referent*.

Of the elements that support the comparison with reference knowledge, in the international corpus as a whole, most are *quasi-referent*. In the Mexican corpus, the exchanges on the price of drinking water provide a large number of elements of this nature, with some discussions raising questions relevant to the learning of economic concepts, such as the role of production factors in determining costs, a first approach to the notion of profit, or the establishment of a link between scarcity and the price of a commodity. The very frequent element corresponding to the concept of virtual water is also *quasi-referent*. The elements, which are very frequent in the American and French corpora, "not all geographical areas have the same water resources" and "there are countries that are richer than others" are also *quasi-referent*. This predominance of *quasi-referent* elements (among the non-doxal ones) goes against the Bachelardian principle[6] that knowledge is built against common opinion, and constitutes and argument for conceptualizing a continuity between knowledge from common experience and scientific knowledge.

Finally, some elements of knowledge-belief, on the other hand, are rather *misleading*. For example, in Mexican schools, the following elements were identified: "water is going to become the new currency" (an obstacle to the understanding of what money is beyond its exchange function), "we have earned the money we need to buy water through our work or our genius" (a caricature of the use of added value whose destination would be limited to final consumption). In the same way, out of the 64 *non-doxal* elements of the American corpus, 10 are not *quasi-referent*. Some of them correspond to past scientific truths or controversies, which, if not updated, may hinder the understanding of current knowledge on these subjects. This is the case, for example, of "there is a climate change, but we cannot say if it is a warming", which is *erroneous*. The comparison of certain elements to reference knowledge is more delicate. For example, "the rich pay much more in taxes than the poor" will not be evaluated in the same way depending on whether one reasons in absolute or relative terms. Among the *non-doxal* descriptive items used by French students, 16 are not *quasi-*

---

[6]Reference to the epistemological work of Gaston Bachelard, for whom profane and scientific knowledge oppose: "A scientific experience [...] contradicts the common experience." (1967 [1938], p. 10).

referent. Some of them are *wrong, for* example "Dubai is in India". But most of them are only likely to hinder the understanding of this knowledge, they are *misleading*. Moreover, some elements used in the French corpus, according to the didactic contexts, rather orientates towards reference knowledge or rather against it. Thus, "one can produce energy" is *quasi-referent* from a technological and industrial point of view, because one indeed speaks of "energy production", to name the process of transformation of one type of energy into another, but one of the bases of physics is precisely the law of conservation of energy, according to which one never creates energy *ex nihilo*. In the context of fundamental physics, it would therefore be rather *wrong*, going against the first law of thermodynamics.

The final criterion is the sources of the items mobilized as "facts." Students in all schools use a variety of sources, but rarely make them explicit. Sometimes linguistic markers are available to infer implicit sources, such as the use of phrases like "it seems that", the conditional tense, etc. Sometimes, no specific sources are directly deducible from students' discourse, and several can likely be hypothesized.

Some of the elements used by the Mexican students come from direct or indirect experience, especially with the description of examples that the students witness. This is often the case for "the rich will have more access to water" and "water resources are decreasing". One occurrence seems to "force" this appeal to experience as a source of authority to support an argument, namely the idea that the climate is changing, presented as an obvious fact that the students would experience directly ("no ves", *you don't see*). We find the exploitation of direct or indirect experience in the American corpus, with descriptions of testimonies ("we can replace meat with other foods", "we need water to live"). In the French corpus, it is mainly the elements linked to the consumption of drinking water that are based on direct or indirect experience: mention of daily practices that consume more or less water, common indicators of its potability such as its color, definition of dry toilets, vital need for water.

Only one statement in the Mexican public school explicitly refers to academic learning, through the use of the teacher's words, in a strategy of argumentation by authority. For example, Donovan, during the small group discussion on OQ1, states:

DON   (...) no se acuerdan lo que dijo la maestra flaviol porque se va para el agua y que luego va a haber guerra (don't you remember what the teacher flaviol said because it is going to water and soon there will be war)

In the other Mexican school, the possibility of using less water is mentioned with explicit reference to the environmental education activities carried out at school: visit to the water museum, viewing of a film on ecology. A third way of referring to school knowledge corresponds to the use of "learnt" vocabulary to mention an idea. For example, a Mexican student uses the term "ecosystem" to argue about the complexity of the consequences of climate change. In the French corpus, we also find several uses of "learnt" vocabulary ("atom", "molecule", "capitalist", "consumer society", etc.). In the American and French corpora, elements that seem to come from school knowledge were also identified by their correspondence to the curriculum ("the Earth is mainly covered by seas", "the quantity of water on Earth is stable", the level of world population, "there are countries that are richer than others", "water is made up of molecules that are themselves made up of atoms", "there is water vapour in the air", "not all geographical areas have the same water resources").

In Mexico, students sometimes openly refer to previous slides to justify their statements. However, by default, it can be considered that whenever information contained in previous questions is mentioned, these can be a source, even if not exclusive. All the sources of this type identified for the American and French corpora are thus implicit. This is the main source for the mention of the virtual water mechanism.
Only among American students is there an explicit reference to parental discourse, on two occasions, for "we need meat to be healthy", and "price differences do not always indicate real quality differences".

As for media sources, the film *Slumdog Millionaire* is explicitly mentioned by Mexican students in relation to the definition of dry toilets. It can be assumed, in this context, that the fact that they are used in India comes from the same source. The only explicit reference to a media source in the American corpus corresponds to the existence of filters that allow students to drink their urine. In three cases, the French students explicitly mention media sources: "it is possible to make water from air", "the end of the world is coming", "water resources are diminishing". Moreover, we can infer that the mentions of the Fukushima accident and the economic crisis in Greece are also based on media sources.

Ordinary argumentations as thoses developed by the students in this corpus exploit descriptive logics. They are based on uncontroversed elements, which are recognized by all the participants to a given interaction as facts. These may be more or less scientifically robust, along a knowing-believing continuum. Some of them can easily be compared to reference knowledge, and might favor its learning (then they are called *quasi-referential*) or, on the contrary, make them harder to understand (then they are called *misleading*), or even be clearly *erroneous*. Other elements of knowing-believing, called *doxal*, cannot be compared to reference knowledge, they are rather beliefs. Three other dimensions characterize the descriptive argumentative resources: their source, either explicit or infered in discourse (daily experience, school, media, parents, etc.), their logical level (object/being or relation), and their degree of generality.

In this corpus, school knowledge does not especifically prevails upon descriptive elements of other sources. Nevertheless, the mentioned facts rather deal with relations between beings or objects than with describing them in isolation, which reveals active reasoning. Moreover, when reformulating a fact mentioned in small group to the whole class, the students often turn it to a higher degree of generality. Such phenomenon is consistent with the well-documented didactical process of institutionalization, corresponding to the elaboration of a generic knowledge from a local, situation-related learning.

2.3   The prescriptive: rules and values

Alongside these descriptive elements, and even, in many cases, in conjunction with them, argumentative discourses also relies on prescriptive logic. One reasons by using prescriptive principles as "compasses" to orient one's argumentative statements. I take up Ducrot's metaphor (Anscombre and Ducrot, 1997 [1983]) concerning the argumentative orientation of discourse: speaking to explore the reasons for defending or, on the contrary, for rejecting a proposition would give a direction to the discourse, the subtle traces of which can be highlighted throughout the construction of the argument. The values or norms of behaviour constituting objects of agreement in the sense of Perelman serve as basic materials for this prescriptive logic. To share them is to admit that all proposals that do not (completely) conform to them must *a priori* be set aside in favour of those that respect them (more). This opens up an immense field of possible justifications and refutations for the evaluation of the alternatives under discussion.

2.3.1 Values and rules of behaviour: definitions

Whereas descriptive logic uses elements of discourse that describe beings and objects in the world and their relationships, prescriptive logic establishes value judgments about them. To do this, it uses general norms about how things should be as argumentative resources. It relies on commonly accepted principles which prescribe what one should believe, say, or do.

Among these resources, it is useful to distinguish between values, on the one hand, and rules of behavior, on the other. The former are a matter of axiological judgment based on abstract moral or ethical principles such as freedom, solidarity, merit, responsibility, justice, etc. Such principles work as ideals to be attained, or to which coming closer through the words, acts or beliefs that we wish to see ratified at the end of the argument. The second ones are rather social rules of good conduct, or even institutional procedures acting as silent traditions accepted as obvious.

Values are not always presented as self-evident, and can give rise to justifications that make use of elements of knowledge-belief, on which a value judgment is applied. Indeed, they are prescriptive, and refer to moral or ethical values, used to evaluate arguments for or against alternative answers to a question. They can be stated in the context of an evaluation of a descriptive element of general scope. This is the case of the following statement: "something must be done or else unfortunately the rich will always have more access to water than the poor", which contains a reference to the fact that "the rich will always have more access to water than the poor", other things being equal, and to the principle of social justice.

Some rules of behaviour have the particularity of moving the discussion from the discursive to the metadiscursive level. These are a specific type of *doxal* elements about the rules prescribing the appropriate behavior in the communicative situation. They thus group together considerations belonging to ordinary interactional politeness and/or specific to the argumentative situation (Goffman, 1974; Brown & Levinson, 1988; Plantin, 2018, p. 369); reagarding the interactional context; in particular the didactic contract in force in the classroom (Chevallard, 1992); students' beliefs about what a good argument is ("ordinary argumentative norms", Doury, 2004), and, in particular, a "scientific" or admissible argument in the school context ("epistemic values", Désautels & Larochelle, 1998; Sandoval, 2005). At the interactional level, we may note, for example, how to behave in class, among classmates, in a debate, when sitting at a roundtable, etc. Among the argumentative norms used in metacognitive comments on the activity in progress, we find elements such as "this is not personal", "we do not resolve disagreements by physical or moral violence". Finally, the epistemological rules engaging the way in which robust thinking is accessed may, for example, concern the need to mention the source of unshared information, the veracity of evidence through experience or testimony, the rule of relevance.

As for the inventory of elements of knowledge-belief, I followed a *bottom-up* approach to identify the prescriptive elements used in this corpus of science *cafés*. The inventory carried out is based on an analysis of the debates in small groups and in whole class about the MQ for the 10 cafés of the corpus, as well as those about the three opinion questions for 6 of them and its results are fully available in my thesis (Polo, 2014, p. 282-299). This inventory work followed 4 steps:

1. identification and grouping of similar elements[7];
2. count of the frequency of elements used in each national corpus;
3. exclusion of infrequent or irrelevant principles[8];
4. identification of a core of prescriptive elements common to all schools, and of local specificities in the exploration of the theme regarding the presence or absence of certain norms or values.

2.3.2 Prescriptive elements used by the students

The results of this inventory show the predominance of norms common to all three countries in the corpus as a whole. I present in Table 6 a summary of this inventory in decreasing order of frequency, for each of these two categories (values and rules of behaviour), which I comment on briefly. Rather than trying to invent paraphrases as analytical labels for each prescriptive element, I have generally chosen to designate them by an emblematic occurrence drawn from the students' discourse, and by a number (V1 for "value 1", N1 for behavioral norm 1). However, in order to facilitate the use of this inventory, in addition to or as a replacement for vague or too extensive quotations, I did sometimes make global or partial paraphrases. They appear in italics.

- *Values used by the students of the three countries (22/35, 63 %)*
Out of the 35 values identified, 22 (63%) were common to all the three countries, with 3 to 52 occurrences (see Table 6). Of course, the fact that students used each of these norms in one country does not mean that everyone in the study in that country agreed about them. In fact, some values present in the three national corpora are even fundamentally opposed, such as the idea that one can act on the course of things (V1, 52 occurrences, including 24 in the 2 Mexican schools, 17 in Kenosha and 11 in Lyon) and that one cannot fight against fatality (V9, 22 occurrences including only 3 in Kenosha).

---

[7]Given the complexity of natural language, the statements collected are inherently polyfunctional, and this at various levels. A single utterance has thus sometimes been considered as referring to several different norms (for example, to both a value and a behavioral norm).
[8]Indeed, some principles were then excluded from the analysis because of their low frequency (less than 3 occurrences) or because they were closely related to local contingencies or digressions during the debate. The principles focused on the local context that were left out referred, for example, to the socio-economic situation of the country or to previous activities done in class.

Table 6. Values identified in the 3 national sub-corpora.

Values		Mx	EU	Fr	Tot.
V1	what we do now will break the em you know who has access to water drinking water in the future - *we can act on the course of things*	24	17	11	**52**
V2	science is always the answer	3	42	5	**50**
V3	si podría ser triste (...) terminar (...) que los ricos tendrán el agua que quieran y los pobres no (yes it could be sad (...) to finish (...) that the rich have all the water they want and the poor no) - *social justice*	18	15	6	**39**
V4	on la distribue à des personnes qui en ont besoin\ (we provide it to people who need it\) - *vital needs must be met*	11	19	2	**32**
V5	l'droit à l'eau potable (the right to drinking water)	6	17	9	**32**
V6	hay que ahorrar el agua (we must save wáter)	28	1	3	**32**
V7	we messed it up so we have to fix it ourselves we messed it up so we have to fix it ourselves - *responsibility*	15	13	1	**29**
V8	bah oui l'argent ça r'vient à tous les coups c'est sûr (well yes money comes up every time that's for sure) - *the value of "money"*	7	7	10	**24**
V9	c'est pas nous qui changerions l'monde\ (it is not us who would change the world\) - *we cannot fight against fatality*	7	3	12	**22**
V10	se está acabando el agua y pues qué compras después con el dinero\ (the water is exhausting and then what will you buy afterwards with the money\) - *money is less important than water*	12	5	4	**21**
V11	je sais pas pourquoi on sacrifierait notre hygiène de vie pour l'eau/ (i don't know why we would sacrifice our lifestyle for water/)	2	5	13	**20**
V12	so that it's cost-effective	3	10	3	**16**
V13	ce s'ra la loi du plus fort (it will be the law of the strongest) - *we must overcome the state of nature*	8	1	4	**13**
V14	no funcionaría porque habría personas que no tendrían agua y se morirían\ (it would not work because there would be people who would not have water and they would die) – *every (human) life deserves to be protected*	6	4	?	**10**
V15	el agua nada más va a servir para tí/ (the water will only be used by you/) - *you have to be in solidarity with others*	1	4	4	**9**
V16	j'partage moi\ (I do share\) – *altruism, one must care for others*	3	2	3	**8**
V17	tenemos que guardar a futuro para nuestros hijos\ (we have to keep in the future for our children) - *we have to think about the well-being of future generations*	2	2	4	**8**
V18	the people are not aware like the water like issue – *one must be aware of the water issue*	3	3	1	**7**
V19	pour euh: la terre\ (for uh: the earth\) - *we have to take care of the planet*	1	5	1	**7**
V20	ayudando a casi toda la pro a toda la poblacion de los países hasta del mundo\ (helping almost all the pro all the population of the countries even of the world\) - *solidarity*	2	2	3	**7**
V21	that's racist\ - *antiracism*	1	1	1	**3**
V22	j'ai pas envie d'faire chier mes parents (i don't want to piss my parents off) - *you have to take care of your family*	1	1	1	**3**

In both Mexican schools, there is a relative emphasis on the fact that current actions determine the evolution of the situation (V1, 24 occurrences and V7, 15 occurrences), especially in terms of social justice (V3, 18 occurrences). The worst possible scenario corresponds to the rule of the strongest (V13, 8 occurrences). The importance of water is emphasized, as having to take precedence over the desire for money, for example (V10, 12 occurrences). Finally, the most specific feature of the Mexican corpus is undoubtedly the elevation of saving water into a moral imperative (V6, 28 occurrences), even if this effort diminishes the quality of life (only 2 of the 20 occurrences of V11).

The Kenosha sub-corpus is characterized by a high frequency of the search for good cost-effectiveness (10 out of 16 occurrences of V12), as well as the universal right to drinking water (17 out of 32 occurrences of V5). The latter goes hand in hand with the norm that everyone has the right to have their needs met (V4, 19 occurrences out of 32), and a preoccupation with the situation on a global scale (5 out of 7 occurrences of V19). Most striking in this school, however, is the great confidence in science as the ultimate solution (42 of 50 occurrences of V2).

Finally, the Lyon sub-corpus presents a high frequency of the feeling of powerlessness towards the evolution of the situation, the belief in the fact that one cannot influence the course of things (12 of the 22 occurrences of V9). This can be related to a rather negative view of human nature, according to which people only think about money (10 of 24 occurrences of V8). In addition, quality of life appears as an important value, with frequent mention of the refusal to "sacrifice it for water" (13 of 20 occurrences of V11).

- *Norms used by the students of the three countries (19/26, 73 %)*
    Out of the 26 rules of behavior identified, 19 (73%) are common to all the three countries. They are listed in Table 7. They appeared from 5 to 43 times in the whole corpus.

Table 7: Norms identified in the 3 national sub-corpora.

Norms		Mx	EU	En	Tot.
N1	el realismo (...) lo que yo tomé en cuenta es la situación realism (...) (realism (...) what I took into account was the situation) – *strong version of the Gricean maxim of quality: one must argue on the probable*	11	21	11	**43**
N2	*if one hesitates between two options, one must choose the one that implies the other (argument exploiting causality)*	17	3	16	**36**

58

N3	Counterexamples are good reasons to refute an argument	1	14	16	31
N4	on parle d'l'eau là on parle pas des voitures\ (we speak about water here we do not speak about cars\) – ignoratio quaestionis, Gricean maxim of relation, rule of relevance	11	7	12	30
N5	si podría ser triste (...) terminar (...) que los ricos tendrán el agua que quieran y los pobres no\ (...) si tiene razón (yes it could be sad (...) ending (...) with rich people getting as much water as they want and the poor people won't\ (...) yes you're right) – we can argue about what is desirable	7	15	5	27
N6	est-ce que tous les pays pourront faire ça/ (will all countries be able to do this/) – we must think globally	2	15	9	26
N7	lo que dirá la mayoría\ (what the majority will say\) – the majority is right	10	5	11	26
N8	one must not contradict himself or herself	7	7	9	23
N9	why is it and how is it A/ – any proposition must be justified	17	3	1	21
N10	il suffit pas d'être riche\ (it is not enough to be rich\) – we must consider a necessary AND sufficient cause	7	1	11	19
N11	elle écoute pas\ elle parle elle écoute pas\ (she doesn't listen\ she talks she doesn't listen |) — we have to listen to each other	9	1	9	19
N12	y quién ganó/ (and who won/) – debating is a competitive activity that ends with someone winning over others	12	5	1	18
N13	entonces entonces qué/ (so so what/) – if you reject a proposal, you must have something else to propose	6	5	5	16
N14	il y a plusieurs réponses\ (there are several answers\) — a controversy admits several rational answers	2	7	4	13
N15	it's about water it's not about politics – it's a technical not a political problem	7	3	1	11

N16	cómo le vas a explicar/ (how are you going to explain it/) – *you have to choose an option that you can easily justify*	8	1	1	**10**
N17	you can't conserve water if you don't have water – *our reasoning must be universalizable*	2	3	4	**9**
N18	ça marche pour maintenant mais pour euh à un moment ça marchera plus\ (it works for now but for uh at some point it will not work anymore\) – *we must think in the long run*	2	5	1	**8**
N19	madame c'est quoi la réponse/ (madam what is the answer/) – *this exercise, like any school activity, admits a single correct answer known in advance by the adults*	3	1	1	**5**

The survey of Mexican schools shows an absence (public school) or very low presence (private school) of concern for access to water on a global scale (N6: 2 occurrences), or in the long term (N18: 2 occurrences). The procedural norm that the issue is a technical rather than a political one is used many times (7 of 11 occurrences of N15). The frequent reminder of the need to justify the proposals made (N9: 17 occurrences and N16: 8 occurrences) reflects a concern to carry out the exercise correctly, and, possibly, a stress towards this unusual activity (especially for students in the public school). The debating activity is indeed seen as more competitive than in the other schools, and the search for a "right answer" is more present (12 of 18 occurrences of N12; 3 of 5 occurrences of N18). It is interesting to note that there is a strong presence of two contradictory rules, namely that one should stick to the probable (N1: 11 occurrences), and that it is possible to argue about the desirable (N5: 7 occurrences).

As regards the American sub-corpus (EU), we find a relative importance of the concern to situate the debate at the global level (N6: 15), but also in the long term (N18: 5 of 8 occurrences). A specificity of this school lies in the high frequency of the idea that one should only argue on the probable (N1: 21 of 43 occurrences). However, the rule according to which one can argue about the desirable (N5 : 15 occurrences) is also very much used. As for the Mexican schools, this double observation is a trace of the confrontation of two opposing arguments elaborated by different students. The relatively frequent mention of the controversial nature of the Question (N14: 7 occurrences) shows a good understanding of the exercise.

In the French corpus (Fr), the debate is also well anchored at the global level (N6 : 9 occurrences). The mention of the rule of non-contradiction of sayings and/or sayings and doings is frequently used

(N8 : 9 of the 23 occurrences). Only one occurrence mentions the fact that the problem would be purely technical (N15), which, associated with the relatively frequent explicitation of the controversial nature of the question, may mean that the students from Lyon have well understood its socioscientific status (N14 : 4 occurrences). The frequent reminder of the rule that we must listen to each other when debating (9 of the 19 occurrences of N11) undoubtedly results from the difficulties of concentration in some of these *cafés*.

Beyond these variations in frequency, this survey reveals the predominance of prescriptive principles common to all the three countries: two thirds of them appear in all national corpora (41/61). How can we explain this "omnipresence", which often even includes contradictory principles? This is what we will discuss in the next section.

2.3.3 Arguing with principles[9]

The presence of most of the prescriptive principles in all the three national corpora may results from the development of arguments that are likely to be opposed in a classical way[10] , which travel beyond national borders. I propose here to explore this hypothesis by apprehending these principles in terms of their interactive elaboration. To this end, I have developed a model of the argumentative use of prescriptive principles in dialogue, drawing on Muntigl and Turnbull's

---

[9]An exploratory but more detailed version of the work discussed in this section was presented in an international conference and published in its proceedings: Polo, C., C. Plantin, K. Lund, G. P. Niccolai (2013a), *Cohering Without Converging: Students' Use of Doxa, Norms and Values while Debating about SSI* (Mexico, USA, France). Proceedings of the *European Science Education Association* (ESERA) *Conference*, Nicosia, Cyprus, September 2–7, 2012, pp. 1387–1399.

[10]These arguments would be "classical" in the sense that they would refer to the *argumentative script* of the question, defined in these terms: "The argumentative script attached to a question includes the set of positions, arguments, counter-arguments and refutations put forward by either party when this issue is debated. They are available to any arguer entering the arena and willing to take a position on the issue. The script corresponds to the state of the argumentative question. It may be implemented any number of times, on a wide variety of forums. It pre-exists and informs concrete argumentative discourses. (...) The first task of the interested party is to review the script relevant to the issue he or she wishes to discuss, and then to perform their partition, that is, to organize a discourse which updates and amplifies the argument line they have selected. In other words, the arguer must define and follow their path within the parameters of the script." (Plantin, 2018, "argumentative script").

(1998) work on the conversational structure of ordinary argumentation.

• *An interactional model of argumentative appeal to norms & values*
Muntigl and Turnbull (1998) have highlighted a conversational structure of ordinary argumentation, based on an analysis of the sequences of 3 turns of speech. It can be schematized as follows:

> T1. A Proposal
> T2. B Disagreement with T1 more or less affecting the face of speaker A
> T3. A Support for own T1 proposal or direct challenge to T2

In the first turn (T1), A makes a proposal. In a second turn (T2), B expresses disagreement with this proposal. The authors have highlighted the fact that the third turn of speech (T3), which is attributed to A, can take two different forms, depending on the intensity of the offense associated with the disagreement expressed in T2. The more T2 offends A's face, the more likely A is not to respond directly to it but to prefer to stick to his/her initial proposal. The less T2 offends A's face, the more A will directly discuss the disagreement expressed in T2.

In the same way of analyzing interactional sequences, the conversational dynamics linked to the argumentative use of prescriptive principles can be modeled, as I propose in figure 4.

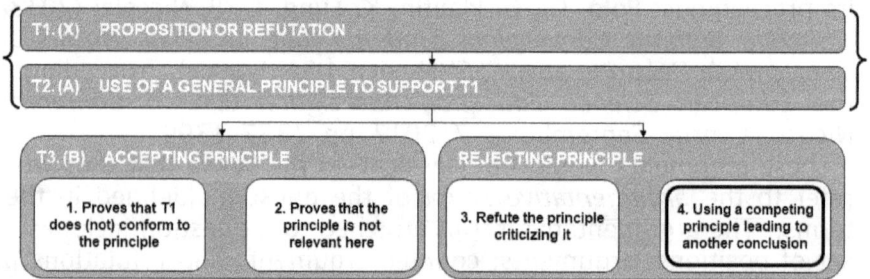

Figure 4. Interactional model of argumentative appeal to principles.

The first turn of speech (T1) corresponds to the formulation of a proposition by a speaker X, who can be either A, B or someone else. In a second turn of speech, A uses a norm or value to support or refute the proposition. In the case where the same person makes a proposition and immediately states a principle to support it (X=A), turns T1 and T2 can in fact be fused, carried out in a single turn. In the next turn (noted here T3), B, who necessarily differs from A, expresses a disagreement with A. In doing so, B can accept (1 or 2) or reject (3 or 4) the prescriptive principle used by A at T2. Four strategies are thus possible: 1) B may discuss the (non-)conformity of the proposition to the principle used (showing its conformity if the appeal to the principle aimed at refuting the proposition; or showing its non-conformity if it aimed at supporting it); 2) B may discuss the relevance of the principle by showing that it does not apply here; 3) B may directly refute the principle used at T2 by criticizing it; or 4) B may use a competing prescriptive principle leading to a contrary evaluation of the proposition. In the latter configuration, the newly used prescriptive principle is presented as superior to the one that was first uttered.

- *Illustrations of the model on examples from the YouTalk corpus*
A discussion about OQ3 among students in the American school provides a typical example of how a proposition is defended by depicting it as actually conforming to the principle used to refute it (type 1 strategy). It is the "think globally" (N6) rule that is used by Gabriel at turn 10:

1.	CAT	[yeah\
2.	ERI	[free drinking water sounds like a horrible idea
3.	ABM	yeah\ or
4.	ERI	cause we [we'll be out of drinking water in like days
5.	GAB	[that's th- [that's c'-
6.	ERI	((laughs))
7.	ABM	that would be: [absolutely
8.	CAT	[yeah\
(.)		
9.	CAT	we [could all like xxx
10.	GAB	[well we're thinking usa again\ we're thinking US again\
(.)		
11.	**ERI**	**[well even from a worldwide standpoint**
12.	GAB	[like free drinking water is terrible yeah\ but the reason you gave me was strictly US and so was my government\
13.	**ERI**	**even from a worldwide standpoint though**

In bold are the utterances that correspond schematically to T3: while the proposal is rejected by Gabriel as being "*strictly US*", Erick tries to present it as valid "*even from a worldwide standpoint*".

A brief exchange from the American corpus also illustrates very well the second strategy of response to the use of a prescriptive principle. As students discuss OQ3, Sean is the only one defending the idea that water should be free. He seeks to rely on N7 that the majority is right, drawing attention to the fact that another table is about to vote for this answer (the A):

1. SEA   okay <((pointing at T6)) they're voting A\>
2. **RIC   no they're not they're lying to you**

His classmate Rick does not directly challenge this rule, but tries to show that it does not apply, by claiming that the students at this table are lying and do not really intend to choose this answer.

Conversely, a clear example of direct refutation of a general principle (type 3 strategy) comes from the French corpus, during a discussion about OQ2. The norm at stake corresponds to the fact of choosing an answer for the whole group, a norm which, in addition to being more or less explicitly used by the students in the discussion, is part of the instructions of the exercise. The rule that we must listen to each other (N11) is also implicitly at stake here.

1. SAM   on n'est pas d'accord\ on n'est pas d'accord\ (we don't agree\ we don't agree\
2. **ASA   bah vous êtes pas d'accord vous allez vous faire [foutre <((en riant)) c'est tout\> (boh: you don't agree you're going to fuck off <((laughing)) that's all\>)**
3. ISA
   [(((pose le E sur le chevalet et le fait tomber)) ([(((puts paper E on the pulpit, which then falls)))

Samira, in turn 1, uses the fact that two of the four students do not agree with the posted answer to justify her decision to lift the card for an alternative answer. However, Asa refuses to consider their opinion, in direct opposition to the two norms Samira relies on.

The type 4 strategy of opposing a superior principle, leading to another conclusion, to the argumentative use of a principle that one rejects, is well illustrated by an excerpt from the discussion about OQ3 during a session hold in the American school.

1. SAB   [and what about (.) family income/ [you need water\

2. PAM   [yeah i also think D too 'cause i don't think like less fortunate people should be (.) punished like you know what i mean like because they don't have money they pay for water they shouldn't (.) [not get water
3. LOU   [yeah:
4. SAB   [xx time it's not their (fault)=

5. **LOU** **=they could like they could overu:se like they could (.) not pay as much and [<(((turning hands)) get more water>&**

6. PAM [and take advantage of that yeah: it's true

7. **LOU** **&take advantage of it\ (.) when like it should be [<((swinging hands)) equal for all people>&**

8. KEL [(((nodding head in the affirmative))

9. LOU &you know what i mean/ 'cause like in like it's their fault that they are (.) poor\ in a way because they could go find a job but they didn't like you know what i mean/

10. PAM yeah

11. **LOU** **like i think it should be equal among everyone** (3.8)

In turn 2, Pamela uses the value of social justice (V3) to support Sabrina's proposal to choose option D (the price of water should depend on family income), which she had already argued with another value (V4: vital needs must be met). Sabrina, in turn 4, strengthens Pamela's argument, supporting the social justice norm (V3) by using the value of merit. Louise then reverses the argumentative orientation of the use of the merit ideal, by showing that option D may lead to a situation that would not conform to it, if the poor took advantage of it to overuse water (turn 5). She is then in a type 1 strategy, and uses the slippery slope argument: the situation would be reversed, with people "*taking advantage*" of their bad behavior, which is not fair according to the ideal of merit. Turn 9 extends this strategy, with a development on individual responsibility in social destiny (V7).

Moreover, Louise does not align herself with the value of social justice. Without opposing it directly, she promotes the use of a competing value, according to a type 4 strategy. Thus, she opposes the principle of absolute equality to the principle of equity, as a frame of reference for what would be fair: "*it should be equal among everyone*" (turns 7 and 11).

Type 4 strategies, which consist in opposing a general principle to counter-argue against another previously used general principle, are particularly interesting to explain the recurrence of a many principles throughout the corpus. Indeed, everything happens as if the use of a principle *orientated* the debate towards another principle that can potentially be opposed to it. The existence of such oppositional scripts in the "argumentative script" (Plantin, 2018, "Argumentative script") associated with a controversy explains why a certain number of principles are frequently used, even in different contexts, namely here the Mexican, American and French schools where we did our investigation.

Prescriptive logics works through the argumentative use of principles, either rules of behaviour (norms) of values. Propositions are then evaluated as more or less acceptable depending whether they do respect these principles or not. If a principle used is consensual, counter-argumentation might either focus on showing that the defended proposition actually conforms to it, or questionning its relevancy in the present context. When there is no prior agreement on the principle used, counter-argumentation might directly refute it, or try to oppose him another principle considered of higher value. Consequently, arguying on the prescriptive plane is likely to activate typical scripts opposing some principles to their usual rival counterparts. Describing such patterns allow for better understanding what is at stake in a debate, revealing the competing underlying worldviews constituting the controversy. Then there is a risk of getting to an unceasing repetition of principles and counter-principles which are given an identity-displaying function. In order to avoid the discussion to get stuck this way, the educators should foster mutual empathetic understanding of the prescriptive basis of adverse argumentative positions.

2.4   The affective: emotional framing of the debate

The aim here is to address how affective logic woks in argumentative discourse: how affects and emotions are used to construct a point of view, and to try to convince others to adopt it. The term "affective logic" allows us to emphasize the fact that this form of emotional thinking is indeed a reasoning process, and cannot be understood in reference to the now obsolete opposition emotion/reason. In this corpus of socioscientific debates between students, it clearly appears that the defense of an argumentative conclusion intrinsically relates to the discursive construction of an emotional position. Indeed, depending on the answer they defend, the students depict (schematize) the problem differently, as well as the alternative answers, which produces an argumentative orientation of the discourse that can be described as emotional framing. More simply, we could say that to describe the affective logic of a debate, for the analyst, is to become capable of considering the words of an arguer in this way: "tell me how you emotionally perceive the problem, and I will guess which alternative you are defending". Of course, we are still talking about the emotional framing of the issue under discussion as it appears in the speaker's discourse, and not about a direct access to his or her thoughts, since any analysis of this type is based on linguistic marks. I first present some fundamental conceptual elements (2.4.1) and then I propose to apply them to a debate sequence from the American corpus (2.4.2).

2.4.1 The affective: conceptual and analytical toolkit

- *Thymical tonality and phasic events*

Plantin (2011) uses the psychological categories of thymical and phasic to analyze the construction of emotions in argumentative discourse. Thymical emotion refers to a relatively stable affective background tonality, a setting in which phasic emotional episodes of greater intensity and shorter duration occur. In the account of an emotional event, the thymical tonality serves as a backdrop for the description of the phasic emotions that take place during this event. If we schematize such a scenario over time (x–axis), the phasic emotions correspond to the "peaks" of intensity (y–axis), according to a more or less bell-shaped curve, and the thymical emotion to the initial level of intensity, generally gradually reached again at the end of the episode (cf. Figure 5). When the curve is drawn upwards, it is an episode of positive emotion and conversely, when it is drawn downwards, the episode is of negative emotion.

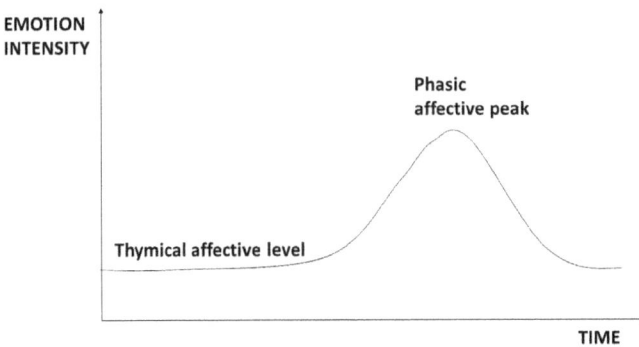

Figure 5: Emotional Episode Curve, reproduced from Plantin (2011, p. 124)

I have chosen to use this distinction between substantive affective tonality and episodic affective variation to describe the operation of affective logic in students' arguments. The thymical, affective background tonality of the debate then refers to a basic emotional framing on which participants generally agree, regardless of their beliefs about the alternatives under discussion. On the other hand, the phasic, episodic, affective variations correspond to the differences in emotional schematization that characterize the competing argumentative positions.

- *The languages of emotion*

On the linguistic plane, the argumentative use of affects and emotions can take more or less explicit and direct forms. Drawing on the distinctions proposed by Plantin (2011), I count four of them. Argued emotions correspond to statements in which emotions are thematized and justified as corresponding to the appropriate affective behaviour. The body of Stéphane Hessel's book Indignez-vous (2010) provides a great example, here of arguing an emotion, here indignation. Said emotions also correspond to an explicit use of affect, but without explicit argumentation. These are statements such as "I am angry" or "We are upset by these attacks". The other two modes of emotion semiotization are implicit and based on inference. If we adopt the stimulus metaphor, the emotions shown are "downstream" signals, implicit but direct, since they are communicative behaviors that express a feeling. For example, "blushing" can express shame. Induced emotions are "upstream" signals: inference is made from the description of a situation which, on the basis of cultural pre-constructs, is associated with a certain emotion. The French expression "avoir la corde au cou" (to have one's neck in a noose) is emblematic of this, expressing a feeling of despair.

- *The seven emotioning topical parameters*

Plantin (2011) has highlighted a number of characteristics that are based on commonly accepted cultural pre-constructs, and that construct emotion in discourse. These "emotioning parameters", taken up by Micheli (2010), constitute the fundamental affective "objects of agreement". These parameters are organized around two structuring axes: intensity (stronger or weaker affect) and valency or agreeableness (more or less pleasant affect). On the axis of agreeableness, the following parameters play a role: proximity to life/death; positive/negative consequences; positive/negative analogies, conformity to norms and values. The parameters contributing to affective intensity are the possibility of control, the possibility of identifying an agent or cause, the social and spatio-temporal distance to the discussed problem. Thus, for example, the cultural pre-construct of preferring life to death constitutes an object of agreement from which a negative emotion is conferred to situations close to death (e.g. to be in deep water), and a positive emotion to situations involving the force of life (e.g. sleeping like a baby). In the same way, a situation presented as close to the argumenters produces a more intense affect than a distant event. We can say that the greater affective sensitivity to things that concern us or are likely to concern us constitutes an object of agreement on which the emotioning parameter of distance to what is being discussed is based.

If these emotional parameters constitute resources for arguing, it is because they are used to construct a specific vision of the question and the alternative answers, by providing an affective framing of the issue. In fact, they make it possible to give an emotional tone to the objects of discourse, and participate in what Grize calls their schematization (1990). By providing an emotional image, they orient the discourse argumentatively towards the conclusion defended. Thus, emotional framing and argumentative positioning are consubstantial (Polo et al. , 2013b). Presenting the unemployed as victims of a global economic system, or as lazy people responsible for their own fate, for example, does not arouse the same emotion towards them (compassion vs. contempt), nor does it imply the same political measure to fight unemployment (stimulus policy vs. control of the unemployed people). Depicting the Question and the alternatives of response according to these two axes determines "topical parameters" (Micheli, 2013, p. 33) that give them a more or less intense, and more or less pleasant, tonality. Analyzing these different parameters thus makes it possible to understand how students frame the debate, and how, by doing so, they orient their discourse towards an argumentative conclusion.

2.4.2 Case study: emotional framing in a final debate

In my dissertation, I showed how this toolkit could inform the debates through three monographs, one chosen for each of the countries in the corpus, from the final debates of the cafés, i.e., those carried out in the whole class about the main question, which follow its discussion in small groups (Polo, 2014, pp. 307–358). The one carried out on the Mexican corpus was chosen for publication in scientific journals (Polo et al. , 2013b, Polo et al. , 2017b) for its emblematic nature and simplicity, related to the fact that, in this debate, opinions polarize between only two options of response. Here, I decipher how affective logic works in a debate that took place in the US school, a case that is easier to grasp than the other two for an English-speaking reader. The question debated is the following:

> In your opinion, in the future, whether a person has access to dinking water will depend on...?
> a) on how rich the person is
> b) on how physically able the person is to live with lower water quality
> c) on efforts made, starting now, to save water by using less and to protect water resources
> d) on where on the globe the person is born
> e) on nature's capacity to adapt to our needs for water
> f) on scientific advances.

Throughout the debate, the students seem rather invested in the task, many of them asking to speak by raising their hands, and not waiting for the facilitator to question them. First, I quickly characterize the thymical tonality of the exchanges. Then, I present the phasic variations linked to the argumentative exploitation of emotions to defend different argumentative conclusions, namely options F, C and D.

- *Thymical tonality: considering the question 'just in case'*

The basic tonality of the debate here contrasts strongly with other debates hold in France or Mexico that frames the issue as a matter of death or life. If the thymical emotion is also rather negative in the present case, its intensity is much lower. Indeed, there is no explicit reference to the risk of death neither to the importance of water to live. Instead, the Question is defined as something abstract, a "very complex issue" (turn 43), or even a "problem" likely to arise "in time" ("so in time it <((hand as getting rid of something)) will (.) be> a problem\", turn 27), and of which it is possible to imagine, at the most, that it could give rise to a "very scary situation" (turn 10).

This relatively low thymical intensity is particularly visible at two levels: the depiction of the involved people and the description of the potential consequences of the problem. Actually, students' discourse build a pretty long thymical distance to the Question: the problem of access to drinking water is defined as not directly concerning the participants to in the debate. When the students use the 1st person, it is to remind how lucky they are to have quality water readily available:

> 3. ABI ours is very good\ so we have easy access to it\

To refer to people who may suffer from a lack of drinking water, the students rather tend to use the third person:

> 7. CAT they move from where they were (...) to like move towards an area with better water source\

Moreover, the places this people come from are not precisely depicted, seeming distant and stranger ("where they were"). The idea that the problem only concerns foreign (for example, in turn 6), is recurrent and never questioned.

> 6. MAR could be born in (like) a (different) country and move somewhere else to have (.) access to: water\

Finally, the people mentioned as being at risk of lacking water have particular characteristics, a priori remote from the students' daily lives. For example, they may be extremely poor:

> 3. ABI if a person is like homeless

In terms of time, these students can only imagine themselves, in a fictional example, as being affected by this problem in several decades:

> 10. NOR   think it's a very scary situation like if we didn't have like (thin) water like how would you feel if you're like in your fourties and you've like kids and you can't give your kid x xx x bath or shower or

Such a temporal localization of the problem contrasts sharply with the very intense thymical tonality of the case study based on a Mexican debate, where a sense of urgency predominates (Polo et al., 2017).
The distance to the Question seems so great that one of the students even explicitly asks the facilitator if she actually believes that they might be affected by this problem one day, to which she does not give an explicit affirmative answer:

> 42. STA   you think we're ever going to run out of water/ in your opinion/
> (.)
> 43. CAT   em i don't know\ it's a very complex issue and like i i'm not sure

As they are not directly concerned, the students should consider this issue due to their moral responsibility as citizens of a world power:

> 44. DIA   it's definitely an issue that we should consider\

Moreover, the description of the potential consequences also contributes to a low emotional tonality: the choice to be made could not affect the extreme of the life–death continuum. At turn 3, although Abigail envisages, in a rather radical turn of phrase, that some people "might not be able to have" water, such a prospect is only described in the conditional tense, except in the case of people on the street ("no water"). Moreover, in the most affected countries, she says, the risk is not being without water at all, but having poor quality water ("water is: ba:d\"). Similarly, at turn 7, the facilitator takes up the idea of water-related migrations not as the search for a water source as an alternative to a total lack of water, but as the search for a "better water source". It is not a question of having or not having water to live, but of living more or less comfortably, with water of more or less good quality.
In terms of quantity, the process is the same: the risk is not to find oneself without water, but to have more or less water. See:

> 19. ROS   it would xx more: available drinking water\ for (those) in the future\

Finally, it seems that what is at stake is not so much the access or not to drinking water, but the ease with which this access will be possible (for example, turn 36: "to get it easier"). One of the criteria evoked to qualify the more or less easy access to water, in addition to the travel time to reach a water source (turn 48), is the cost of its production, in particular in reference to the desalination of sea water:

54. DIA   [&and we still have the problem of digging the salt out of that is [very expensive

One excerpt is emblematic of such thymical tonality:

46. STA   <((shrugging)) just in case\>

DIA   [(((nodding head in the affirmative))
47. CAT   [&we should think about\ and something that as a first world country we don't often think about\ you know you got your <((hand movement as if she was doing it)) faucet turning on> <(((showing where the water would be)) there's your water\>

ABI ((hand up))

CAT   but like in other places
48. STA   yes they go and walk and get it [xx water\

In response to the facilitator's not telling her that this is a problem that directly concerns them (turns 42 and 43), Stacey schematizes the debate activity they are conducting as having a preventive or precautionary function: "just in case" (turn 46). Diana, the other facilitator, agrees with this schematization of the debate anticipate the issue "just in case" the situation changes. Cathy, the facilitator with whom this exchange began, confirms this privileged position of the participants regarding access to drinking water, and gives an explanation linked to the economic and geopolitical power of the country ("as a first world country", turn 47). This element also functions as an argument to get the students interested in the debate, whether they gain from learning about the situation of other countries in the world, or whether they feel responsible for it, as citizens of a very powerful country on the international level. In any case, the students' investment here can be based on a feeling of direct confrontation to the problem at stake.

Still, not all students have the same emotional position. Depending on whether they defend options F, C or D, they orient their discourse emotionally and argumentatively in distinct ways. Such "phasic" variations are described below.

- *Competing emotional schematizations as phasic argumentative resources*

Three main options are competing in this debate: C) access to drinking water in the future will depend on efforts made now to save and preserve water; D) it will depend on where on the planet one was born; and F) it will depend on scientific advances. While option D embodies an emotional position of fatality, rather negatively oriented, but intended to be realistic, the other two options are alternatives to this unpleasant situation, both of which admit the idea of some control over the course of things, allowing for positive evolution. I will now detail these emotional schematizations more precisely, first in terms of intensity (distance to the problem, possibility of control, agency/causality), then, in terms of valency (mainly with conformity to general principles).

Phasic building of the distance to the problem

Regarding the people affected by the risk of a lack of drinking water, the supporters of option D tend to refer to people far away from them, whereas the supporters of options C and F are more likely to include themselves as being involved in the situation. Abigail, defending the idea that access to water in the future will depend mainly on where one was born (D), uses the third person ("*they*", "*a person*", turns 3 and 64) to evoke serious situations on the life–death axis, whether it is a question of not being *able to get* water at all ("no *water*", turn 3), or of suffering a deterioration of one's immune system ("*their immune system can become depleted*", turn 64). She also sometimes uses the first person plural, but associated with a more abstract and less negatively oriented risk ("*a problem*", turn 27); or with a possible solution (turn 52). Jim, who is hesitating between several answers, uses the second person, when he describes option D as interesting: "*if you live in a place that's more populated with water than other places then you'll probably be able to get it easier*" (turn 36). However, this is a general, abstract "*you*", which *does* not particularly involve the people present and which *is*, here too, associated with a smaller risk: not having or not having water, but getting it more or less easily. The spatial framing of the problem reinforces this emotional distance: it affects "some countries", but not ours ("*some countries*" turn 3), or "some places" (turn 36), but not here. When Abigail uses the pronoun "we", in turn 52, she refers in a very general way to humanity, and not to the territory on which the participants in the debate live, mentioning the possibility of exploiting the melting of the polar ice to get drinking water.

In contrast, the supporters of option C (*access to water in the future will depend on the efforts made to save and preserve it*), directly present themselves as concerned by the problem of access to water. The use of the second person is more concrete here: it is about oneself or one's children (turn 10); and alternates with the first person plural (turn 30: "*we'll have enough*"). Moreover, the consequences described are more intense emotionnally, affective one's own body with the mention of hygiene (turn 10) and water to drink (turn 19). The possibility of a lack of water is even explicitly considered (turn 30). Supporters of option C never make the spatial schematization of the problem explicit, which, combined with the use of the first two persons, does not exclude the territory where the debate is taking place.

In terms of emotional distance to the problem, the supporters of option F (*access to water will depend on scientific advances*) develop an intermediate position as compared to students defending options C and D. They refer to people suffering from a lack of water in the third person (turns 6 and 48), but they use the first person plural (turns 22–24, 42), or a general "you" (turn 43) expressing a commitment to solidarity and a choice to feel concerned by the problem. The difficulties of a few to access water are presented as a challenge for humanity, to which we should all seek to respond, for example through the desalination of sea water (turns 22–24). This distinction between people actually affected by the problem of access to water and people who feel concerned and involved to solve it is uptaken at turn 44, when Stacey asks, as if she needed to be reassured, whether they might actually run out of water one day. Here, the "we" refers to the concrete community to which the people in the room belong, and the answer is fairly obvious: nobody feels directly threatened. In terms of spatial distance to the problem, supporters of option F are on a similarly intermediate position: they describe the areas really affected as far away ("*different country*", "*other places*", turns 6 and 48), but the possible places of solution are not specified, and do not exclude the American territory, especially the coasts, with the mention of using sea water (turn 22–24).

In terms of temporal distance from the problem, the specificity of the emotional schematization of the problem by the supporters of option C (it is *the effort made* that counts) is particularly visible. Indeed, all their interventions explicitly mention the fact that the situation could be problematic in the future, but is not currently ("*in your fourties*"; "*in the future*"). This is consistent with the option chosen: only the existence of a time frame for action can make it possible to believe that the efforts made could pay off and positively impact the evolution of access to water. On the contrary, supporters of option D present the problem as already existing in the present, notably with the use of the present tense to describe general truth (turns 3 and 36). When future tense is used (turns 27, 36, 64), it is to describe processes that take place in time, but that are not specific to a future situation that would be different from the present one. For example, they mention that drinking poor quality water can affect people's immune systems (turn

64: "*if a: physic'lly able person was drinking like em: not very good water/ then their immune system can [become depleted*]"). This temporal framing of the problem, as resulting from a phenomena independent of the time in which one is situated, is constitutive of a fatalistic position such as option D (*access to water in the future will depend on where one is born on the planet*).

As for the supporters of option F (*access to water will depend on scientific advances*), while they describe the problem in the present tense, they use many action verbs that imply a sufficient time frame to react to the situation. Stacey's phrase in turn 42 is emblematic of this type of discursive construction: "*to adapt to the water we need*". The reference to movement, used here in the literal sense (turns 6 and 48), can constitute a good analytical metaphor to characterize this emotional position: the observation of current needs is not associated with a getting stuck in an ineluctable situation, but motivates positive change, "progress". We are here almost in an argument based on faith: since we need it, we will eventually find solutions (turn 33). The position of the supporters of F (scientific advances) regarding the distance to the problem is closer to the thymical tonality (rather negative but not very intense), than the one of the students defending the other answers. However, the study of the temporal distance already reveals a specificity of the position of the supporters of F regarding the possibility of controlling the evolution of the situation. This other dimension, along with the determination or not of a cause or an agent responsible for the situation, constitute another pillar of construction of the students' emotional position, on the axis of intensity.

Table 8 lists, for each of these 3 options, the contributions of the students to the debate that participate in the discursive construction of their distance to the problem of access to water.

Table 8: Distance to problem built by students supporting options C, D or F.

Distance to the problem of accessing water – Supporters of C	
10 NOR	if we didn't have like (thin) water like how **would you** feel if you're like **in your fourties** and you've like kids and **you can't give your kid** x xx x bath or shower or
19 ROS	so we could now if we start to use less xx and it would xx more: available drinking water\ for **(those) in the future**
30 ROS	maybe **for the future we**'ll have enough
Distance to the problem – Supporters of D	
3 ABI	**in some countries** the water is: ba:d\ (...) **they might not be** able to get it and ours is very good\ so we have easy access to it\ and: some of the other options like em A i

	don't think i– em: it should d– depend on that\ because if **a person is like homeless** then that **means** no water\
27 ABI	i don't think that would help **us** because yeah if we do start like saving water now (.) it's still gonna: like we're still gonna be using it\ so **in time** it <((hand chase gesture)) will (.) be> a problem\
36 JIM	if **you** live in **a place that's more populated with water than other places** then **you'**ll probably be able to get it easier
52 ABI	how **we** can't just use the water from like the polar icecaps that are melting\
64 ABI	because like (.) e: if **a: physic'lly able person** was drinking like em: not very good water/ then **their** immune system can [become depleted and] then **THEY** would need [the more purified water\
Distance to the problem – Supporters of F	
6 MAR	**someone** could be born in (like) **a (different) country** and move somewhere else to have (.) access to: water\
22–24 FRA	**we** (will/really) nee:d a more effective way a less exp-expensive way to: get the salt out of (.) salt water\ so:\
33 MAR	like they would **end up finding** ways like **you** could probably get water and xx xx\
42 STA	<((shrugs)) to adapt to the water> **we** need/
44 STA	you think **we'll ever** run out of water/
48 STA	yes **they go and walk** and get it [xx water\ (turn 47: "**in other places**")

Phasic variation among the possibility of being in control and the determination of causality/agency

Table 9 highlights the contributions to the debate that participate in different emotional schematizations in terms of the possibility to control the evolution of access to water, and the attribution (or not) of responsibility to a particular cause or agent.

Table 9: Potential control and causality/agency determination regarding changes in access to water, depending on the by defended option

	Potential control over the evolution of access to water – supporters of C
10 NOR	**people** really don't think about like when they're doing stuff\ like saving like x xx xx they xxbly (haven't had)\ they don't necessarily like to think what would be **the action that could really like (outload) x backwards or another like the simple like stuff** hairs'cream make–up x shampoo on water and em
19 ROS	**what we do now will break** the em you know who has access to water drinking water in the future\. because if we abuse you know like using water in a way we are now it would it should xxx **probably**\ so we could now if we start to use less xx and it would xx more: available drinking water\ for (those) in the future
30 ROS	**if we develop a system** in which we can continue to save water **instead of just (.) trying once in saving a quarter free (renove it)** to really make it in the future so: that'd be: like also on scientific development creates different techniques and advances in order to: save er: water\ and protect it **maybe for the future** we'll have enough
36 JIM	if **we can start** saving water now and like what rose said you know (lay on) to continue saving again\ then **that could help** too\

Potential control over the evolution of access to water – supporters of D	
3 ABI	in some countries **the water is: ba:d\ so: they might not be able** to get it (...) if a person is like homeless then **that means** no water\
27 ABI	**i don't think that would help** us because yeah if we do start like saving water now (.) it's still gonna: like **we're still gonna be using it\ so in time** it <((hand chase gesture)) will (.) be> a problem\
52 ABI	**why we have to save water** but then em: i don't get how **we can't just use the water from like the polar icecaps** that are melting\
36 JIM	**if you live in a place that's more populated** with water than other places then you'll probably be able to get it easier
Potential control over the evolution of access to water – supporters of F	
6 MAR	**someone could** be born in (like) a (different) country and **move** somewhere else to have (.) access to: water\
22–24 FRA	salt water\ i think it's gonna be (.) **it's gonna have to be** that thing that **we use** the most\
33 MAR	**science** always xx like **they would end up finding ways** like you could probably get water and xx xx\
42 STA	<((shrugs)) to adapt to the water> we need/ **but (.) people use water** for nothing (in place in) or what/

Indeed, these dimensions are intrisincally linked: presenting the situation as controllable often goes hand in hand with identifying a cause on which it is possible to act. They influence both the emotional intensity, a greater possibility of control being generally associated with a lower emotional intensity, and the nature of the emotions depicted. For instance, you cannot revolt against fatality, whereas you can be outraged about a crime. I do not have the means to go deeper into the delicate field of the exact denomination of the emotions at stake. There is no need for such labelling analysis, which is rarely consensual, to adress my research question, namely to understand how students use emotions as argumentative resources. In this perspective, describing the phasic variation in terms of relative position on the axes of intensity and valency is sufficient to characterize the alternative rival emotional positions constructed co-substantially to the defense of competing argumentative conclusions.

For the proponents of option C (*current efforts will determine future access to water*), the possibility of control over the evolution of the situation is not in doubt. The latter will not evolve spontaneously, but is likely to improve or, on the contrary, to deteriorate, as a result of human (in)action. This potentiality is mainly expressed in the conditional tense ("*could*", turns 10 and 36), and with the use of softening adverbs ("*really*", turn 10; "*probably*", turn 19; "*maybe*", turn 30). All the causes of the evolution of the situation mentioned are human agents, namely "people" in general ("*people*", turn 10), most frequently designated by the use of the first person plural (turns 19, 30, 36). This self-attribution of collective responsibility functions as a call to action, and has the advantage of denouncing the bad habits of the debaters without personal attacks. Conviction then appears as relating on identification with a shared community, aware of the limitations of its way of life, than on the recognition of individual guilt. Such an emotional schematization has contrasting effects in terms of the emotional intensity of the debate. The feeling of control offers room of improvement, and then tends to decrease the intensity of the thymical tonality. Conversely, depicting the debaters' community as responsible for taking action to solve the issue constitutes an emotional intensification, by "bringing the problem closer" to the present people.

Students who support the idea that access to water in the future will depend on where one is born (D) see little or no possibility of control over the situation. Their position on this point is radically opposed to that of people defending option C. In turn 27, Abigail makes this disagreement explicit, about the effectiveness of efforts to save water: "*i don't think that would help*". In particular, the disagreement is about the feasibility of reducing the use of soft water. Indeed, she believes that in the long term, such effort cannot be sufficient. This goes hand in hand with the attribution of the current situation, and the lack of water, not to human agents, but to natural causes beyond human control. While Norah sees our wasteful use of water as an important factor in explaining water supply levels (turn 10), the proponents of option D point instead to the unequal endowment/dotation of water between territories (turn 36), or to the poor quality of some water sources (turn 3). Even when students defending D refer to the quality of water, they present it as a fact, without mentioning any process that would be responsible for its pollution: "the water is bad" (turn 3). From then on, the only option is to accept one's fate of having the good or back luck of living in a the place which is more or less rich in water. In this fatalistic vision, only spontaneous natural changes might contribute to an evolution of the situation, as the melting of the polar ice (turn 52). Such a position has two phasic effects: 1) it raises the emotional intensity of the problem, because of the impossibility of controlling it; and 2) it makes it more acceptable, because it evolves independently from human will.

The proponents of option F (access to water will depend on scientific advances) rather consider that control over the evolution of access to water is possible, unlike the proponents of D, and as do the proponents of C. However, they differ from the students defending C in terms of agency. Stacey, in turn 42, is skeptical about people reducing their water consumption ("*but people use water*"). In this respect, the emotional schematization is aligned with that of the supporters of option D. Thus, at the individual level, the only possibility of action considered is to move to a territory with a better water supply (turn 6), which constitutes a recognition of the weight of fatality regarding the initial water resources. But this turn also functions as a counter-argument to option D: the actual defense of option F, namely the determining role of science in the evolution of access to water, relies on the staging of a different agent in this debate. Thus, turns 22, 24 and 33 refer to "science" ("*science*") and to scientists ("*they*") as the main actors on whom change depends. The modal expressions here shows an unlimited belief in their ability to solve the water problem: "*science always...*" (turn 33); "*it's gonna have to be*" (turns 22-24). In the emotional position elaborated by the supporters of option F, the possibility of control, which is limited for everyone, is devolved to the scientists, who are presented as very trustful. The prospect of being able to rely on people more powerful than oneself to solve the problem constitutes, among the three positions described here, the one that most diminishes the emotional intensity of the thymical tonality. From a rather negative background tonality, it even shifts the debate to positive emotions.

An analysis of the construction of these different emotional positions on the axis of valency makes it possible to specify this tendency of the supporters of option F to bring the debate into pleasant emotions. Several dimensions play an important role in the construction of students' emotional positions on such axis: the use of an emotional lexicon, the dramatization of the consequences of the different options of answer, especially on the life-death axis, and the reference to shared norms and laws.

Phasic variation on the valency axis: emotional lexicon
The use of an emotional lexicon is part of the emotional construction of the problem and the different options. By "emotional lexicon", I mean both the use of the typical vocabulary of emotion, and the use of terms with emotional connotations. Here, I build on the perspective of Ortony, Clore & Foss (1987) who propose three different types of emotional orientation of words : focus on the affective, the behavioral or the cognitive dimension of the emotion. In the studied debate, there are few cases of use of the classical vocabulary of emotion, and most of the emotional lexicon refers to terms with emotional connotation or orientation. The table 10 below lists them, in the contributions of the supporters of the three options C, D and F.

The only occurrence of an emotional classical term, "*scary*", referring to fear, occurs among the advocates of option C (*access to water will depend on our efforts*) and describes a situation of lack of water (turn 10). This situation corresponds to the impossibility of performing a very positively oriented emotional act: giving one's children water to wash. The term "kids" is indeed clearly life–oriented, and therefore positive in terms of emotional schematization. Here, combined with a negative predicate, it intensifies the negative perspective of running out of water.

Table 10: Emotional Lexicon of the supporters of the 3 options C, D and F.

Emotional lexicon		
Tour	Option	Excerpt
10 NOR	C	it's a very **scary** situation like if we didn't have like (thin) water like how would you feel if you're like in your fourties and you've like **kids** and you can't **give your kid** x xx x bath or shower or
19 ROS	C	if we **abuse** you know like using water in a way we are now it would it should xxx probably\ so we could now if we start to use less xx and it would xx more: available **drinking** water\ for (those) in the future\
30 ROS	C	we can continue to **save** water instead of just (.) trying once in **saving** a quarter free (renove it) (...) scientific **development** creates different techniques and **advances** in order to: **save** er: water\ and **protect** it
36 JIM	C	em also C: like (.) if we can start **saving** water now (...) then that could **help** too\
3 ABI	D	in some countries the water is: ba:d\ so: (...) ours is very **good**\ so we have **easy** access to it\ (...) if a person is like **homeless** then that means no water\
27 ABI	D	i don't think that would **help** us because yeah if we do start like **saving** water now (.) it's still gonna: like we're still gonna be using it\ so in time it <((hand as getting rid of something)) will (.) be> a **problem**\
36 JIM	D	if you live in a place that's more populated with water than other places then you'll probably be able to get it **easier**
64 ABI	D	if a: **physic'lly able** person was drinking like em: **not very good** water/ then their immune system can [become **depleted** and] then THEY would **need** [the more purified water\

24 FRA	F	cause like we (will/really) **nee:d** a more effective way
40 STA	F	to adapt to the water> we **need/**
42 STA	F	we're ever **run out of** water
52 ABI	F	like they were saying to like **save** water (...) why we have to **save** water

As the transitive verb "*to give,*" it is a positively emotional term in most of its uses, which is evidenced by the definition in Webster's online dictionary (accessed April 6th, 2023)[11]: one usually gives things that are nice to receive (gifts, parties, rights, medicine; blessings; fruit; one's trust; attention, money, one's word, etc).

1. to make a present of <give a doll to a child>
2. a: to grant or bestow by formal action <the law gives citizens the right to vote>
   b: to accord or yield to another <gave him her confidence>
3. a: to put into the possession of another for his or her use <gave me his phone number>
   b (1): to administer as a sacrament (2): to administer as a medicine

   c: to commit to another as a trust or responsibility and usually for an expressed reas
   d: to transfer from one's authority or custody < the sheriff *gave* the prisoner to the warden>
   e: to execute and deliver <all employees must give bond>

   f: to convey to another <give them my regards>
4. a : to offer to the action of another : proffer <gave her his hand>

   b : to yield (oneself) to a man in sexual intercourse
5. a : to present in public performance <give a concert>

   b : to present to view or observation <gave the signal to start>
6. to provide by way of entertainment <give a party>
7. to propose as a toast

   (...)
8. d : to attribute in thought or utterance : ascribe <gave the credit to you>

---

[11] As for the value of using the dictionary to identify the emotional lexicon, see the very clear justification for this approach in Plantin (2012).

9. a : to yield as a product, consequence, or effect : produce <cows give milk> <84 divided by 12 gives 7>
    b : to bring forth : bear
10. a : to yield possession of by way of exchange : pay

    b : to dispose of for a price : sell

    (...)
11. d: to award by formal verdict <judgment was given against the plaintiff>
12. to offer for consideration, acceptance, or use <gives no reason for his absence>
    (...)
13. b: to offer as appropriate or due especially to something higher or more worthy <gave his spirit to God>
    c: to apply freely or fully: devote <gave themselves to their work>

    d : to offer as a pledge <I give you my word>
14. a: to cause one to have or receive <mountains always gave him pleasure>
    (...)
15. a: to allow one to have or take <give me time>

    (...)
16. to care to the extent of <didn't give a hoot>

In Norah's intervention, the evocation of fear and the impossibility to keep on with a daily, very positive situation, contribute to present the future, if nothing changes, as very negative. It is only a short step from there to asserting the need for change. Rose takes this step when, in turn 19, she links this schematization of the problem to the choice of option C, introducing it as a solution to avoid such a fate, and to have more drinking water in the future. In addition, Rose uses the emotional term "to *abuse*," which refers to shame, to describe the situation where nothing would change and no effort would be made, a negative emotion that adds to those previously mobilized to show that it is undesirable.

1. a: to put to a **wrong** or **improper** use <abuse a privilege>.
        b: to use **excessively** <abuse alcohol>; also: to use without medical justification <abusing painkillers>
2. obsolete : **deceive**
3. too use so as to **injure** or **damage** : **maltreat** <abused his wife>

4. to **attack** in words: revile <verbally abused the referee>

The other two interventions, in turns 30 and 36, instead feature a positive emotional lexicon to depict the efforts made and their effects. The efforts are described with a recurrence of the verb "*to save*" (4 occurrences), one occurrence of the verb "*to protect*", and, for the specific type of effort that constitutes scientific research, with that of "*development*". All of them are very positively-orientated words. See notably the definition of the verb '*to save*' below.

1. a : to **deliver from sin**

    b : to **rescue or deliver from danger or harm**

    c : to **preserve or guard from injury, destruction, or loss**

    d : to store (data) in a computer or on a storage device (such as a CD or flash drive)
2. a : to put aside as a store or reserve : accumulate <saving money for emergencies>
    b : to spend less by <save 25 percent>
3. a : to **make unnecessary** : to avoid <it saves an hour's driving>.

    b (1) : to **keep from being lost** to an opponent (2) : to prevent an opponent from scoring or winning
4. to maintain, **preserve** <save appearances>

Thus, even though the verb "*to save*" is used here in the sense of its definition in Webster, it is less emotionally neutral than "to store". I is positively connoted as allowing one to escape from something negative (sin, danger, pain, loss, destruction, etc.), namely, in this context, the risk of running out of water. Using the term in combination with the verb "*to protect*" (see below) strengthens such connotation.

1. a : to **cover or shield from exposure, injury, damage, or** destruction : guard   b : **defend** <protect the goal>
2. to maintain the status or **integrity** of especially through financial or legal guarantees (...)

Finally, the term "*development*" refers to a positive evolution, as shown by the examples given in its definition in the dictionary, very life-oriented, with the notions of progress and fulfillment (muscular or professional development, stage of development):

1. the act, process, or result of developing <development of new ideas> <an interesting development>
2. the state of being developed <a project in development>

3. a tract of land that has been made available or usable : a developed tract of land especially : one with houses built on it

This is consistent with the fact that these students also depict the results of the efforts considered in option C in a positive way, with the use of the terms "*advances*" and "*help*". The term "*advance*", which is close to "*development*", refers to an improvement, with the metaphor of movement to signify an increase in value (sense 3 of the dictionary), or a move closer to the targeted goal.

1. a moving forward <halted the enemy's advances>
2. a: progress in development <mistaking material advance for spiritual enrichment – H. J. Laski>
   b: a progressive step: improvement <an advance in medical technique>.
3. a rise in price, value, or amount <The workers won wage advances.>
4. a first step or approach made <her attitude discouraged all advances> (...)

As it comes to the verb "*to help*", it explicitly referes as an action that makes a situation more pleasant, more emotionally enjoyable (sense 2 of the dictionary). It is interesting to note that such definition also evokes the metaphor of advancement (sense 3), and, in its archaic sense, the verb "*to save*".

1. to give assistance or support to (someone) : to provide (someone) with something that is useful or necessary in achieving an end <He helps the children with their homework.> <Can you help me get this jar open?>
2. a: to **make more pleasant or bearable**: improve, relieve <bright curtains will help the room> <took an aspirin to help her headache>
   b archaic : **rescue, save** <Help us from famine>
3. a : to be of use to : benefit <will do anything to help their cause>
   b : to **further the advancement** of : promote <could help negotiations>
4. a : to **change for the better**   (...)

In turn 27, Abigail, who is defending option D, expresses her disagreement with option C, using the emotional term "*to help*" with a negation. While she aligns herself with the use of "*save*" and the presentation of the efforts made as a positive process, she refuses the description of its result as sufficient, and uses, to define the situation even after such efforts, the negative term "*problem*". A problem is, in fact, defined, at least in one of its meanings (2b), as a cause of negative emotions (perplexity, stress, vexation):

1.   a : a question raised for inquiry, consideration, or solution
        (...)
2.   a : an intricate unsettled question
        b : a **source of perplexity, distress, or vexation**
        c: difficulty in understanding or accepting <I have a problem
with your saying that>

Similarly, in turn 64, Abigail rejects option B as likely to transform a
positive state of physical ability, clearly oriented towards life, into a
state of need ("*need*") associated with depletion ("*depleted*"), terms
more oriented towards death. On the other hand, when the supporters
of D defend their option (*access to water will depend on the place of
birth*), they insist on the dichotomy between places well endowed with
water and places poorly endowed with water (turns 3 and 36), the
former corresponding to the places where the participants live, and the
latter to remote places. The emotive term corresponding to this
description of the situation, in addition to the qualifiers of good/bad
characterizing the water resources, is the adjective "*easy*", or its
comparative of superiority "*easier*". Here is the definition:

1. a: causing or involving **little difficulty or discomfort** <within easy
    reach>
    (...)
2. b : not steep or abrupt <easy slopes>.
    c: **not difficult to endure** or undergo <an easy penalty>.
    d: readily taken advantage of <an easy target for takeovers> <an
    easy mark for con men>
    e (1) : readily available <easy pickings> (2) : plentiful in supply at
    low or declining interest rates <easy money> (3) : less in demand
    and usually lower in price <bonds were easier>
    f: **pleasant** <easy listening>
    (...)
3. a : marked by **peace and comfort** <the easy life of a courtier>
    b: **not hurried or strenuous** <an easy pace>.
4. a: **free from pain, annoyance, or anxiety** <did all she could to
    make him easier>
    (...)
5. a : giving **ease, comfort, or relaxation**
    b : **not burdensome or straitened** <bought on easy terms>.
    (...)
    d: marked by ready facility <an easy flowing style>.
    e: felt or attained to readily, naturally, and spontaneously <an easy
    smile>

In many of its uses, "easy" thus refers at least to the absence of
negative emotions (stress, anxiety, pain, boredom), and at most to the
presence of a certain number of positive emotions (calm, comfort,

pleasure). The situation thus described is all the more acceptable as it corresponds to a rather pleasant state of affairs for the debaters. Indeed, there is no mention of the negative consequences involved for people living in territories with more limited resources, whereas there is a discussion on the negative effects of other options. For example, in turn 3, Abigail gives a radical negative description of the consequence of option A (*access to water will depend on income*), for the poorest people. In addition to the death-orientation of the expression "*no water*," mobilizing the image of "homelessness*" schematizes those affected as deserving pity, thanks to the emotional connotation of the term. It is defined as having no "home" ("*having no home or permanent place of residence*," "homeless," *Webster* online, April 6th 2023), with the term "home" referring beyond the living building to the sense of well-being and comfort associated with familiar surroundings:

1. a : one's place of residence : domicile <has been away from home for two weeks>
   b: house <several homes for sale in the area>
2. the social unit formed by a **family** living together (...) <comes from a loving home>
3. a : a familiar or usual setting : congenial environment; also : the focus of one's domestic attention <home is where the heart is>
   (...)
4. a: a place of origin <salmon returning to their home to spawn>; also: one's own country <having troubles at home and abroad>
   b: headquarters <home of the dance company>.
5. an establishment providing residence and **care** for people with special needs <homes for the elderly>.
   (...)
– at home
   1: **relaxed and comfortable : at ease** <felt completely at home on the stage>
   2: **in harmony** with the surroundings
   3: on familiar ground: knowledgeable <teachers at home in their subject fields>

On the contrary, the supporters of option F (*access to water will depend on scientific advances*) share with those of option C a rather negative emotional characterization of the situation as it is and risks evolving if no change is made. The risk underlined is that of a lack, clearly oriented towards death by the non-satisfaction of vital needs ("*need*"; "*run out of*", turns 24, 40, 42). This schematization works, here too, as a call to action to avoid such an extremity. However, option C is rejected (turn 52), and option F is presented as a more pleasant alternative, the "scientific advances" being more positive than the "efforts" to be made.

Arguments by consequences: position on the life–death axis

If the study of the lexicon already suggests different emotional positioning of the problem on the life–death axis by the students defending different options, considering the consequences of the different options discussed clarify these orientations. The following table lists the interventions corresponding to such an emotional orientation by the consequences.

Table 11: Consequence orientation by advocated options

Emotional orientation through consequences		
Turn	Option	Excerpt
10 NOR	C	in your fourties and you've like kids and you can't give your kid x xx x bath or shower or
19 ROS	C	it would xx more: available drinking water\ for (those) in the future
30 ROS	C	maybe for the future we'll have enough
36 JIM	C	then that could help too
3 ABI	D	they might not be able to get it (...) so we have easy access to it (...) if a person is like homeless then that means no water\
27 ABI	D	em well i was gonna respond to them saying C\ and i don't think it would be that (.) or i don't think that would help us because yeah if we do start like saving water now (.) it's still gonna: like we're still gonna be using it\ so in time it <((hand as getting rid of something)) will (.) be> a problem\
52 ABI	D	i don't get how we can't just use the water from like the polar icecaps that are melting\
64 ABI	D	then THEY would need [the more purified water\
36 JIM	D	then you gonna probably gonna be able to get it easier\
6 MAR	F	move somewhere else to have (.) access to: water\
33 MAR	F	you could probably get water and xx xx\

Without any change, the consequences described by the supporters of options C (*access to water will depend on the efforts made*) and F (*it will depend on scientific advances*) of a spontaneous evolution of the situation are very strongly oriented towards death. It is a question of not being able to provide basic hygiene conditions for one's children (turn 10), of not having enough water (turn 30), or even of not having access to water (turn 6). Such a negative description appears only once among the students defending option D (*access to water will depend on the place of birth*), in turn 3, with the risk of not having water, but this is presented as concerning only far people (either geographically or socially). The same person, Abigail, presents the spontaneous evolution of the situation differently, in turn 52, by imagining that the problem of access to water will be solved naturally thanks to the polar ice melting, making human inaction much less problematic. Similarly, Jim's intervention in turn 36 depicts the consequence of no change as maintaining more or less easy access to water, which is far less death-oriented than the radical perspective of no access to water.

The consequences of option B (*access to water will depend on the physical ability to live with water of lower quality*) are only considered in turn 64, where this option is rejected. They are presented as a new problematic state, rather than as a state where a solution has been found. Thus, the most physically resistant people would experience a deterioration of their immune system that would lead them to require better quality water from now on. Clearly death-oriented, the consequence thus described lends a very negative emotional tone to option B.

As for the consequences of the efforts involved in option C, they are presented differently by the proponents and by the supporters of option D. The proponents of C only mention very positive, life-oriented consequences: access to enough water, more available drinking water, or a more pleasant situation that has arisen as a result of these efforts (turns 19, 30, 36). In contrast, in turn 27, Abigail, advocating for the competing option D, describes such efforts as insufficient in the long run, and leading to a "problem".

Only one instance, in turn 33, directly evokes a consequence of option F, which would lead to the promotion of scientific research to facilitate access to water: "*you could probably get water*". Thus, one of its supporters, Marlene, presents it as positively oriented, towards life, with, as a main consequence, the likeliness that everyone would have access to water.

Using shared principles to determine options' acceptability
On the axis of valency, this attribution of more or less positive emotional tones to the different options does not only involve the use of a specific lexicon and arguments by consequences. In their discourse, the students also construct them by referring to a certain number of norms and laws. A proposition will be all the more positive on the valency axis (and, or credible) if it is presented as conforming the general principles considered as shared common ground, either values (noted v), procedural norms (about how to behave in the debate, noted n), or laws (noted l). Tables 12, 13 and 14 inventory this use of norms and laws to attribute different emotional tones to options C, D and F, and its argumentative effects.

89

It's mainly its advocates who discuss the compliance of option C with shared principles, together with the supporters of the competing option D. The supporters of C justify it by emphasizing 5 values (turns 10, 19, 30, 36). The first (v1) corresponds to a negative judgment about a general law, namely that people waste water (l1). In line with this condemnation of such a practice, and in accordance with a principle set up as a fundamental norm, that of taking care of water (v5), efforts should be made to save and conserve water (option C). Option C, moreover, is presented as consistent with the belief that one can and should make a difference (v2), that there is a universal right to access clean water (v3), and the duty to provide water to one's children and, more generally, to be responsible towards future generations (v4, v4+). Moreover, option C obeys the law that if you save something, you have more in the future (l2).

Table 12: Option C's Compliance with shared principles according to advocated options

Discussion on compliance with principles – option C (access to water will depend on our efforts)				
Turn	Option	Excerpt	Principle(s)	Argumentative effects
10 NOR	C	people really don't think about like when they're doing stuff\ like saving like x xx xx they xxbly (haven't had)\ they don't necessarily like think what would be the action that could really like (outload) x backwards or another like the simple like stuff like hairs'cream make-up x shampoo on water and em i think it's a very scary situation like if we didn't have like (thin) water like how would you feel if you're like in your fourties and you've like kids and you can't give your kid x xx x bath or shower or	v1: (l1) = evil v2: one can act on the course of things v3: right to water v4: provide for one's children	Justification of C by a principle, examples or it and then, *a contrario*, by the negative consequences of not choosing it
19 ROS	C	what we do now will break the em you know who has access to water drinking water in the future\. because if we abuse you know like using water in a way we are now it would it should xxx probably\ so we could now if we start to use less xx and it would xx more: available drinking water\ for (those) in the future	v2+: we can act on the course of things v3+ v5: taking care of water l2: savings > more water v4+: think about future generations	Support for C with rephrasing and addition of its positive consequences
27 ABI	D	i don't think that would help us because yeah if we do start like saving water now (.) it's still gonna: like we're still gonna be using it\ so in time it <((hand as getting rid of something)) will (.) be> a problem\	v4+': (l2)=not realistic in the long term p1: long term solution	Opposition to C based on doubts about its feasibility

90

30 ROS	C	if we develop a system in which we can continue to save water instead of just (.) trying once in saving a quarter free (renove it) to really make it in the future so:	l2+ p1+ v4++	Counter-argument for C taking into account the long term
36 JIM	C	em also C: like (.) if we can start saving water now and like what rose said you know (lay on) to continue saving again\ then that could help too\	l2++	Support for C by rephrasing the last argument
52 ABI	D	they were saying to like save water\ but then like they're also saying that the polar icecaps are melting\ so: like i don't get why: (...) why we have to save water but then em: i don't get how we can't just use the water from like the polar icecaps that are melting\	l2': melting > more water v6: nature's resilience	Opposition to C by challenging the premise of water scarcity

The proponents of option D position themselves as opposing Option C, and evaluate it negatively, with reference to several principles. In response to the mobilization of v4+ as a basis for the choice of option C, concerning responsibility towards future generations, they frame such option, and, in particular, the underlying economics (l2), as unrealistic in the long term (v4+', turn 27). In doing so, the principle of having to choose an option that constitutes a long–term solution (n1) appears. In turn 30, this norm is taken up by Rose, who, in an argumentative reversal, uses it to defend option C (n1+), specifying how the efforts made can be fruitful through time (l2+), and effectively influence access to water in the future (v4++). Finally, in turn 52, option C is compared to option D. Saving water would not be essential to increase the amount of water available in the future (l2), which would increase spontaneously resulting from the polar ice melting (l2'). Therefore, the efforts involved in option C would be unnecessary and would even conflict with the fundamental belief in the resilience of nature (v6).

In a similar pattern, the use of principles to evaluate option D relies on its' advocates' discourse (positive judgment) and interventions by supporters of option F (negative judgment).

Table 13: Option D's compliance with shared principles according to advocated options

Discussion on compliance with principles – option D (the place of birth)				
Turn	Option	Excerpt	Principle(s)	Argumentative effects
3 ABI	D	in some countries the water is: ba:d\ so: they might not be able to get it and ours is very good\ so we have easy access to it\	I3: unequal endowment; I4: water here; n2: thinking globally	D as an obvious choice, based on the local example
6 MAR	F	i would say it wouldn't be D:\ because someone could be born in (like) a (different) country and move somewhere else to have (.) access to: water\	I5: migration for water	Counter-example opposing D
36 JIM	D	or even D like if you live in a place that's more populated with water than other places then you'll probably be able to get it easier\	I3+; I4: easy access on its territory	Rephrasing D as an obvious choice
50 STA	F	yes they go and walk and get it [xx water\	I3++: places where access is less easy	Concession to D

Staying on their *ethos of* realism, the proponents of option D mostly use norms to defend their conclusion. Two items are primarily exploited: the observation of the unequal endowment of water resources across the territories (I3), and the observation of the easy local access to water on their shared territory (I4). The first point is conceded by Stacey, who defends option F, in turn 50, but is not considered sufficient to justify the choice of option D. However, another law is opposed by Marlene, in turn 6: the existence of migratory flows that can allow people to escape a fate that would be entirely determined by their place of birth (I5).

Finally, Option D is presented as conforming to the procedural norm of thinking about the problem as a whole, on a global scale (p2, turn 3), an approach not disputed.

Only students supporting option F (access to water will depend on scientific advances) assess its compliance with shared principles (Table 33).

Table 14: Option F's compliance with shared principles

Discussion on compliance with principles – F (access to water will depend on scientific advances)				
Turn	Option	Excerpt	Principles	Argumentative effects
22-24 FRA	F	salt water\ i think it's gonna be (.) it's gonna have to be that thing that we use the most\ 'cause like we (will/really) nee:d a more effective way a less exp- expensive way to: get the salt out of (.) salt water\ so:\	v3: right to water v7: faith in science I5: possible desalination v8: cost-effectiveness	F justified by a technique presented as the obvious solution, meeting shared ideals
30 ROS	F	like also on scientific development creates different techniques and advances in order to: save er: water\ and protect it maybe for the future we'll have enough	v3+ v7+ v5	Support for F by proclaiming a general faith in technology
33 MAR	F	science always xx like they would end up finding ways like you could probably get water and xx xx\	v3++ v7++	Rephrasing of the previous justification

Students mainly use values, two of which are also used to defend option C, namely the universal right to access water (v3), and the duty to take care of water (v5). Franck introduces in turns 22–24 the value of cost effectiveness: the best solution is the cheapest one (v8). It relates to the example of desalination, its possibility working as an argument for option F (I5). Finally, it is mainly the faith in science (v7) that specificities the position of the supporters of option F as differing from those of option C. Present in each of their interventions such a value constitutes a key part of the view on the problem and, more broadly, of their underlying worldview.

During the YouTalk, the students argue framing the debate in a way that restrains the possible subsequent turns, allowing them to defend their opinion. An important part of such framing is emotional. The emotional position built *orientates* towards the corresponding argumentative conclusion. The diverging options selected to interpret the problem have direct consequences on the possible ways to deal with it. If, *'in common speech, building an argumentative conclusion, is building an emotional position'* (Plantin 2011: 5), it seems here that building up an emotional position is choosing an argumentative conclusion. Such results show that it is impossible, in authentic discourse, to separate reason from emotion. Indeed, in practice, any instruction aiming at excluding affective reasoning to focus on a supposed more objective emotion-free position would turn to taking a side without confessing it. This is because emotionally framing the problem, even on a low-intensity basis, is never argumentatively neutral. For instance, displaying a 'realistic' perspective and promoting a proposition by appealing to the principle that argumentation should only be based on facts, can be a great strategy to counter-argue opponents who would depict it with a negative affective tonality referring to common values. On the contrary, welcoming the affective framing of argumentative discourse as constitutive of its very nature allows for fully apprehending it and for understanding why complex issues can be controversial in society.

III. Reasoning together: conditions and indicators of a successful discussion

Anyone trying to bring people to explore together a complex issue quickly realizes the weight of "social" or "relational" factors in the success of such project. People talk about the "fruitful working climate" of a discussion in which everyone ends up satisfied, with the feeling of having improved. On the other hand, we deplore the "ego wars" or the lack of personal investment in failures to reach true collective cognitive advancement. Sometimes, obvious interpersonal differences help explain such failures. But often oneself, either participating in the discussion or facilitating it, is not sure what is behind, why the collective reasoning occurs... or not. The data I studied is no exception to this rule. When watching the debates between students, one quickly senses that their problem is not only to think about the question at stake, but also to manage what it implies to think collectively: a certain number of relational and identity-related issues are invited to the table. I therefore sought to understand what was at stake in this "social" dimension, to open this black box that is sometimes presented as a key element of any socioscientific debate without being clearly defined. This led me to compare the results from different research fields, to study the quality of collective reasoning as it appears in the students' discourse, and to formulate hypotheses to explain that this quality can vary from one debate to another, sometimes for the same group of people. As I sought to deepen my understanding of the role of emotions in argumentative interactions, the link between affective logic and this social dimension of exchanges became obvious. This relation is explored in the second part of this chapter (3.2), where I present a first model of the social and cognitive functions of emotions in collective reasoning.

3.1 Quality of the argumentative work in group

Although it is not yet clear how to assess the quality of a student debate about a socioscientific controversy, the literature points to four key elements: knowledge about the topic being discussed (Lewis & Leach, 2006), understanding of the interdisciplinary and controversial nature of the issue (Driver, et al. 1996), students' epistemic values, i.e. their conception of what knowledge is (Désautels & Larochelle, 1998, Sandoval, 2005), and the quality of social interactions during debates (Albe, 2006). However, the literature on the didactics of socioscientific controversies has only rarely explored this last element, which I have chosen to call the social dimension of argumentative work in groups. I therefore had to combine analytical and interpretive tools from other disciplines in order to try to identify what a quality group argument could be at the social level, and to propose a method to apprehend it. After having shared a synthesis of the readings that inspired my approach (3.1.1), I present here the five indicators of the quality of group reasoning that I have developed (3.1.2), and their application to concrete cases (3.1.3, 3.1.4, 3.1.5). Finally, I build from this confrontation between theory and data a toolbox to evaluate collective argumentation: the Multi-Level Group talk Analysis Grid (ADiGM) (3.1.6).

3.1.1 Reasoning together as a communicative act

When trying to understand how interactions in a group of debating students work, it is useful to draw on two traditions in linguistic research: interactional linguistics and argumentation studies.

- *"Saving the face" as a structuring dynamic of human interactions*
Interactional linguistics investigates the "grammar of interaction", i.e. in how conversations work. While these conversations may *a priori* give an impression of disorganization, they actually follow such organizational principles that it is possible to predict some aspects of the continuation of the exchanges, from the analysis of the first turns of speech. One of the undisputed results of interactional linguistics is to emphasize the existence of a system of *politeness* which largely structures verbal interactions. Here, *politeness* must be understood as everything that someone does, in interaction, in order to preserve his or her *face* or that of others, *face* standing for the positive social value attributed to a person (Goffman, 1974, p. 15). We thus speak of 'facework' to refer to all the discursive productions aiming, in the course of the interaction, at meeting the rules of politeness: avoiding speech acts that could threaten someone's *face* or, when they cannot be avoided, trying to soften them with other speech acts that restore the *face* (Brown & Levinson, 1988). For example, a refusal will be associated with an apology, for instance, if you cannot give someone a cigarette: "*Sorry*, I don't smoke". Such *facework* dynamics constraint the discursive sequence: the production of a given speech act will partially determine the continuation of the dialogue by directing it towards one of the few possible *polite* reactions. The speech acts that respect the rules of linguistic politeness are therefore considered as *preferred* to those that deviate from them: they are more likely to happen. Notably, empirical studies show that, in ordinary conversation, agreement is *preferred* to disagreement, the latter being considered as a threat to the *face* (Pomerantz, 1984).

- *Specificities of the argumentative interaction*

While such speech act ordinarily appears as linguistic *impoliteness*, the explicit expression of disagreement is central to the definition of argumentation. Indeed, in everyday life, interactions tend to be more focused on interpersonal relations than on the topics discussed: the issue is to create, maintain or strengthen the social link rather than to solve a cognitive problem. However, when the topic of the conversation is dramatic enough to the speakers, they may make a disagreement explicit, and progressively develop an argumentative sequence.

Nevertheless, in the literature of argumentation studies, this behavior then is not considered as *impolite*. Rather than breaking the rules of *politeness*, the situation is better understood as a shift towards another system of *politeness*, specific to the argumentative context, which allows and encourages the expression and justification of disagreements. Thus, in an argumentative context, it would be allowed, and even expected, to criticize another's arguments, which would then not be a threat to one's *face* (Plantin, 2018, 'Politeness'). Sometimes, this system of *politeness* specific to argumentative situations is even depicted as an inverted politeness: criticizing others' arguments is then accepted as a necessary act to understand what the disagreement is based on and to go deeper into the addressed issue. Eventually, such criticism appears to as more *polite* than an unmotivated disagreement, which does not allow the other to counter-argue and tends to block the discussion.

- *School debate about a SSC*

The context of debating about a socioscientific controversy at school also affects how students interact with each other. Indeed, such interdisciplinary issues involve knowledge of different epistemic levels, and are subject to diverse subjective assessments, thus giving rise to controversies which a simple problem–solving approach is not sufficient to apprehend. In order to explore them, students may argue in a way that differs from argumentation aimed at finding the answer to a question that would admit an exact and unique answer. They may experience strong disagreements among themselves, even persisting as they deepen their knowledge on the topic, because they actually reflect different worldviews. Therefore, the teaching of SSC necessarily relies on several pedagogical aims, in an extensive conception of learning, going beyond the simple acuisition of knowledge, and integrating issues such as: development of critical thinking, ability to discriminate knowledge coming from different sources of information, ability to deepen the space of debate in order to perceive the complex dimension of the problem, and use of argumentative skills.

- *Collaborative learning and constructive group talk*

Research on collaborative learning has specifically studied group argumentation as a means of deepening understanding of various phenomena and as a basis for learning, both at school, at the workplace, or in informal education contexts. Such approach implies considering the group as a real cognitive unit, distinct from the sum of its individual members, and questions what promotes quality reasoning at the group level (Stahl, 2006). Most of the work carried out from this perspective is based on analyses of the characteristics of the discussion within the group.

These research questions have led to the definition of types of classroom talk considered to be of pedagogical interest. Mercer and Wegerif were among the first authors trying to define a form of discourse of particular interest to small group learning in the context of the mathematics classroom: "*exploratory talk*" (Mercer, 1996; Wegerif & Mercer, 1997). According to them, it is distinct from two other types of small group talk of lower educational value: "*cumulative talk*" and "*disputational talk*".

Let's take a closer look at these three categories, which have greatly inspired my work on the social dimension of group argumentation. In its very definition, the exploratory talk is considered as the explicitation of thought allowing its collective appropriation and deepening:

> First it is talk in which partners present ideas as clearly and as explicitly as necessary for them to become shared and jointly evaluated. Second, it is talk in which partners reason together - problems are jointly analyzed, possible explanations are compared, joint decisions are reached. From an observer's point of view, their reasoning is visible in the talk . (Mercer, 1996, p. 363).

This concept of exploratory talk has been adapted for the analysis of SSC. The authors then no longer put much emphasis on consensus building. Instead, they ask to which extend the students are able to understand alternative points of view, and to construct dialogical arguments, taking into account potential counter-arguments (Albe, 2006; Lewis & Leach, 2006).

However, Mercer notes that students often fail to engage into *exploratory talk*, and fall into what he sees as two pitfalls: *cumulative* of *disputationnal* talk. *Disputational* talk is characterized by persistent disagreement and individual decision-making, and is embodied in short, assertion-counter-assertion exchanges, without justifications. Conversely, in *cumulative* talk, speakers co-elaborate the discussion in an uncritical way, through frequent repetitions and confirmations (Mercer, op. cit. , p. 369).

It is interesting to note that this typology also corresponds to three different sources of social recognition for the students:

> In exploratory talk (…) the 'yes' of cumulative talk and the 'no' of disputational talk move almost instantly (…) to the construction of different kinds of self-identity. The first constructs and maintains self-identity as in solidarity with the physically present group while the second constructs individual self-identity in opposition to others. Exploratory talk, on the other hand, does not appear to imply or require a specific form of identity commitment. By engaging in exploratory talk, participants maintain a psychological detachment both from themselves as individuals and from the group. (…) In exploratory talk, then, one ultimately identifies neither with one's own self nor with a group but rather with the dialogue. (Wegerif & Mercer, 1997, pp. 54-56).

In order to foster engagement into *exploratory* talk, *ground rules* should be established and followed in the classroom: sharing information in the group, encouraging all group members to speak, consideration and respect for all ideas, making clear one's reasons for thinking what one is proposing, acceptance of criticism, making criticisms and alternatives explicit and negotiating them, discussion of all alternatives before any decision is made, seeking consensus for any action or decision, and holding the whole group accountable for the decisions made (Fernández et al. 2002, p. 43; Wegerif et al. 2004, p. 144; Mercer & Sams, 2006, p. 512). These principles converge with the four indicators proposed by Asterhan to identify what she calls 'productive argumentation': a general willingness to listen to and critically examine different ideas, a willingness to make concessions in response to competing arguments, an atmosphere of collaboration and mutual respect, and a perception of the activity as a competition between ideas and not between individuals (Asterhan, 2013, p. 254). I interpret these rules as ways to create the conditions necessary for each member of the group to have their search for social recognition met in a way that promotes *exploratory* talk. Indeed, in my opinion, there is a specific source of recognition, linked to a way of projecting one's identity in the interaction, which characterizes the *exploratory* talk, and deserves to be described beyond the vague idea of "identification with the dialogue". This is one of the points I have sought to clarify in my work.

Moreover, while authors agree that some forms of interaction and discussion are more useful for learning than others, the problem remains of operationalizing these broad concepts to assess the quality of authentic dialogues. The *Cambridge Educational Dialogue Research* group (CEDiR), to which Mercer belongs, has recently been working on this issue of operationalizing the study of discourse in educational contexts, in collaboration with a Mexican team from the *Universidad Nacional Autónoma de México* (UNAM). This led them to develop a broad analytical grid, the *Scheme for Educational Dialogue Analysis* (SEDA), which aims to study any form of dialogue-based pedagogical situation, and to question the quality of associated talk (Hennessy et al. , 2016). Some of the 33 indicators in this grid are of course related to the characteristics of *exploratory talk*, even if they are not specifically aimed at studying talk between students.

I share with this group the concern to find indicators to identify, in authentic interactions, marks of collective reasoning. However, I draw more attention to the interactional context, and I believe that it is difficult to build a single grid that would be relevant for any pedagogical situation. I have therefore chosen to focus on group talk in symmetrical interactions, between students of the same age, in the absence of a teacher and any other adult. In the corpus studied here, these are the small group discussions preceding the whole class debates. My approach also differs from that of SEDA as it is more descriptive than normative. Of course, as an educator, I am also concerned with fostering quality group reasoning. However, I begin by trying to better understand the actual practices of students when they argue with each other, whether or not they correspond to the types of talk considered of highest educational value, rather than imposing an ideal model of discussion on pedagogical situations. Moreover, to avoid reducing group reasoning to a sum of individual contributions, I propose to study collectively constructed discourse sequences rather than individual turns of speech as SEDA does. Finally, as a linguist, I am aware that utterances can serve several purposes at the same time: they are polyfunctional. In this respects, SEDA seems limited to me since it only allows to attribute one function to a turn of speech (technically, this grid relies on mutually exclusive analytic codes).

3.1.2 Assessing the quality of group talk

The concern to assess the quality of group talk led me to highlight 5 indicators of its propensity to foster collective reasoning. The quality of small group discussions, from the point of view of their cognitive advancement, can thus be evaluated according to the levels of *justification of opinions*, *topical alignment*, *critical examination*, *cooperative decision-making*, and *dialogic strenghtening of arguments*.

• *Five indicators of group talk quality*

The question guiding the study of the first indicator, *justification of opinions*, is "are propositions and rebuttals justified?" This criterion is fully satisfied when any statement accepting or rejecting a proposition

100

is supported by an element of justification. This is a characteristic feature of exploratory talk, which assumes that "challenges are justified and alternative hypotheses are offered " (Mercer, 1996, p. 369). In a group, the justification of a proposition or rebuttal can, of course, be produced directly by the person who made the assertion, or by someone else. In addition, the justification may be spontaneously associated with the assertion, or it may occur after another person has expressed skepticism about the proposition or rebuttal, and/or asked for it to be argued.

The second indicator, *topical alignment*, refers to the fact that the students are effectively engaged, at the cognitive level, in exploring the ideas contributed by each member of the group (op. cit. , p. 369). In concrete terms, it is a matter of looking at whether the students, through their contributions, are indeed developing a discourse based on the argumentative content of the previous speech turns. The linguistic markers constituting traces of this *topical alignment* are typically verbal or gestural repetitions of referents associated with the argumentative content under discussion.

The third indicator, the existence of a dynamic of *critical examination* of the arguments presented in the discussion, corresponds to asking whether the students evaluate the ideas contributed by each member of the group. It is not just a matter of being critical, but of examining each idea, whether or not it is *ultimately* selected, and to assess it critically, but constructively (op. cit., p. 369). Indeed, *exploratory talk* requires all arguments to be truly analyzed and compared (Mercer, 1996, p. 363).

The fourth indicator of the quality of group talk is the level of cooperation in decision–making. Are all group members considered when making a decision for the group? In this corpus, when there is a real cooperative dynamic, students seek agreement from all group members on the choice of an option for the collective vote that closes the discussion. However, reaching such a "joint decision" (Mercer, 1996, p. 363) does not necessarily mean that there is a consensus. For example, it is common for groups of students to choose to break the rules of the exercise by presenting two responses, in order to ensure that all opinions of the group are represented, or to have the opportunity to explain how their position is intermediate between the two posted responses.

The fifth and last indicator of the quality of the group talk, the *dialogic strengthening of arguments,* highlights how the exchanges in the group enrich the argumentation developed about the issue under discussion. It is a question of identifying, in the argumentative strenghtening of individual contributions, traces of collective reasoning. Do individual contributions to the class debate integrate the arguments and counter–arguments presented earlier by other group members, either to build on these ideas or, on the contrary, to argue against them? This indicator is associated with the question of the extent to which the group as a whole feels "responsible" and has "appropriated" the decision made and the arguments that support it (op. cit., p. 371). In the context of the YouTalk, this refers to the fact that, during the whole-class debate following the collective vote, any member of the group can formulate any of the arguments developed previously during the group discussion, and not only those he or she had initially brought. This last indicator has the advantage of showing how learning takes place through three interrelated levels. Thus, it shows that quality individual contributions, presenting complex dialogic arguments, result from prior contact with counter or supporting discourses developed during the group discussion. The level of the small group of students and the whole class are also related, since the challenge of participating in the class debate is an essential motivation for the group discussion. This transfer of some of the arguments developed in the small group to the classroom space can even be seen, to some extent, as a form of *institutionalization* of the stronger arguments built in the group.

A discussion is considered to belong to *exploratory talk* when meeting these five criteria. When some indicators are missing or only partially achieved, it allows the analyst to specify the definitions of *disputational* and *cumulative talk*. Before illustrating the how such indicators work on authentic case studies, it is worth giving a brief quantitative overview of their occurrences in the analyzed corpus.

- *Inventory of the types of group talk in the corpus*
I used these 5 criteria to identify which of the 76 small group discussions on opinion questions in the corpus fell into *exploratory talk, disputational talk*, and *cumulative talk*. Such inventory revealed that a great number of discussions could not easily be classified into one or another of these three categories, as they seemed to alternate between moments of *exploration* and/or *dispute;* or *exploratory talk* and/or *cumulative talk*. This led me to identify 43 *hybrid* cases, distinct from the *typical* cases that fall into only one type of group talk. Table 15 presents the results of this global inventory.

Table 15. Quantitative overview of the inventory of the types of group talk in the YouTalk corpus.

	Exploratory	Cumulative	Disputational	Hybrid	Total
American public school	4	6	0	8	18
French public school	4	0	3	11	18

102

Mexican public school	3	3	0	14	**20**
Mexican private school	8	2	0	10	**20**
Total	**19**	**11**	**3**	**43**	**76**

These 5 indicators made it possible to identify cases of *exploratory talk* in each of the schools visited, and therefore appear not to be very sensitive to language and culture. This point is essential when one seeks to highlight the characteristics of reasoning itself without prejudging the superiority of one cultural rhetorical style over another. In other words, no matter whether one debates with a "Mexican style", "American style" or "French style" (if the national reference is ever relevant to determine the way of arguing), he or she can experience *exploratory talk*. On the other hand, I found no cases of *cumulative talk* in the French data, which are the only ones to present, conversely, typical cases of *disputational talk.* However, among the hybrid cases listed, I could find, locally, moments of *cumulative* and *disputational talk* in each of the three the national corpora.

3.1.3   A case of successful argumentation: *exploratory talk*

Let's see how these indicators work in order to understand a *typical* case of *exploratory talk*: the discussion between Pamela, Louise, Kelly and Sabrina on the third opinion question in the American school. It is about choosing how the price of water should be determined (see fig. 6).

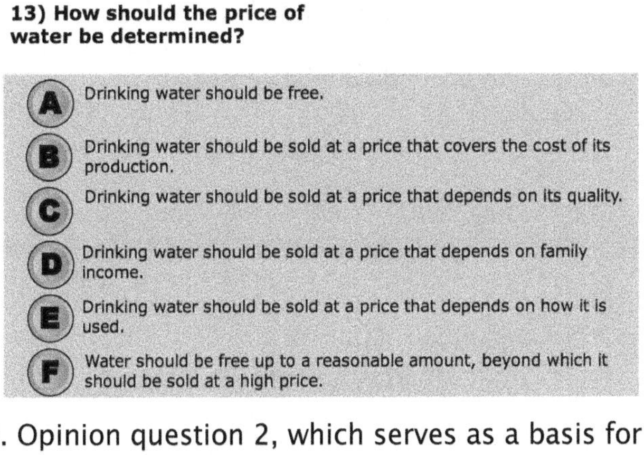

13) How should the price of water be determined?

A Drinking water should be free.

B Drinking water should be sold at a price that covers the cost of its production.

C Drinking water should be sold at a price that depends on its quality.

D Drinking water should be sold at a price that depends on family income.

E Drinking water should be sold at a price that depends on how it is used.

F Water should be free up to a reasonable amount, beyond which it should be sold at a high price.

Figure 6. Opinion question 2, which serves as a basis for Pamela, Sabrina, Kelly and Louise's *exploratory* dialogue.

The discussion is polarized between two options: the price of drinking water should depend on its quality (C) or it should vary according to family income (D). Their dialogue is transcribed below, including, when considered by te groupe, the parallel speech of the facilitator (ANI), addressed to the whole class.

1. LOU   er: i think it should be priced by its quality because if [you'd have better quality it's just more work to like produce it\
2. KEL   [((nods head in the affirmative, looking at Louise))
3. KEL   em: there's more [production <((turning hands)) for it to>
4. LOU   [((turning hands))
5. LOU   yeah\
6. KEL   yeah\
7. KEL   [em:
8. SAB   [and what about (.) family income/ [you need water\
9. PAM   [yeah i also think D too 'cause like i don't think like less fortunate people should be (.) punished like you know what i mean like because they don't have money they pay for water they shouldn't (.) [not get water
10. LOU
[yeah:
11. SAB
[xx time it's not their (fault)=
12. LOU   =they could like they could overu:se like they could (.) not pay as much and [<((turning hands)) get more water>&
13. PAM   [and take advantage of that yeah: it's true
14. LOU   &take advantage of it\ (.) when like it should be [<((swinging hands)) equal for all people>&
15. KEL   [((nods head in the affirmative))
16. LOU   &you know what i mean/ 'cause like in like it's their fault that they are (.) poor\ in a way because they could go find a job but they didn't like you know what i mean/
17. PAM   yeah
18. LOU   like i think it should be equal among everyone\
(3.8)
19. KEL   er:
20. LOU   er: i would say C but what are you guys [saying/
21. SAB   [what did you x the quality that was
22. LOU   like [in a xxx water
23. KEL   [that's bad water or [the water
24. LOU   [xxx water is like more expensive than our like gross water which is more expensive than like
25. LOU   [((shrugs))

104

```
26. KEL     [it's because like it's like [processed more and like
27. LOU                        [xxx water\
28. SAB                        [((nods head in the affirmative))
29. LOU     it's processed more [and
30. SAB                        [yeah\ i think it's either C or D\
31. LOU     it actually takes work to go like get it and find out xx
32. MO1     okay so: [if you guys actually wanna pull up your letter
now/ let's get started
33. SAB                   [maybe C AND D\ 'cause like
34. PAM     just pu– just put C:\
35. SAB     ((taking card C))
36. KEL     put D\ no put C
37. MO1     you guys put your letters [up xx\
38. SAB                               [what/
39. KEL     i don't know:\   (0.8)
40. SAB     [we should put
41. PAM     [just put C and i'll explain like why we think D too\
42. KEL     yeah:\
43. SAB     <((putting card C)) well i'm putting C\>
44. LOU     C\
```

Let's start the analysis with the first indicator, the *justification of opinions*. The two options, when suggested, are immediately justified. In turn 1, Louise gives as a reason for choosing option C (a price for water according to its quality): "*better quality it's just more* work". Similarly, in turn 8, Sabrina justifies option D (a price according to family income): "*you need water*". When she expresses her support for option D, Pamela, in turn 9, also elaborates a justification: "i don't think like less fortunate people should be (.) punished like you know what i mean like because they don't have money they pay for water". Note that such justifications appeal to values: on the one hand, the ideal of merit and fair reward for work (option C); on the other, the right to the satisfaction of basic needs and the ideal of social justice (option D). Moreover, the argument for the latter option reverses the argumentative orientation of the use of the ideal of merit: no punishment is deserved just for belonging to a poor socio–economic group.Students also use justifications when refuting claims. When Louise counter–argues option D, she gives three reasons for doing so:
1) an argument based on the consequences, pointing out the risk of overexploitation of cheap water by some categories of population: "*they could overu:se*"(turn 12);
2) a second inversion of the use of merit, by presenting the poor as actually responsible for their social position: "*it's their fault that they are (.) poor in a way*", (turn 16);
3) a substitution of the concept of absolute equality for that of social equity as an ideal of justice, used and repeated as a motto: "*it should be equal among everyone*" (turns 14 and 18).

Here are also positive the second and third indicators of the quality of group talk, *topical alignment*, and *critical examination* of ideas. For example, when she counter-argues, Louise makes the effot to actually consider and evaluate the rival proposition before going back to her own (indicator 3). Kelly co-constructs with Louise, in turns 1 to 7, a justification of proposition C ("*it's like processed more*") and rephrases it in turn 26, before Louise uses it again (turns 29 and 31). Such a collaboration shows a real circulation of ideas between students. Pamela also behaves according to criteria 2 and 3: she tries to understand the arguments of others and to take part to their elaboration, even if she does not agree with the corresponding proposition. For example, in turn 13, she completes Louise's sentence even though she is counter-arguing her own proposition. Here, her reaction "*it's true*" is ambiguous: it is not clear whether it should be interpreted as a concession or a change of opinion. Starting in turn 21, Sabrina also commits to exploring option C, even though she prefers option D, encouraging Louise and Kelly to detail the reasons why they would choose it. She evaluates their arguments positively in turns 28 and 30, but when it is time to vote, Sabrina, although agreeing to post option C, also wants to add her initial choice: "*maybe C AND D*" (turn 33).

It is then very interesting to observe the collective decision-making process enacted by the four students, despite the time constraint. They actively seek the assent of all group members, conforming to the fourth indicator of *exploratory talk*. Already in turn 20, Louise had explicitly asked the others for their opinion. Sabrina tries to reconcile the divergent views by proposing to display both answers, even though this is contrary to the rules of the exercise. Then Pamela acts as argumentation studies call a *third party* (Plantin, 2018, "Roles"): she temporarily steps back, and away from her own position, to facilitate group collaboration. Concretely, she proposes an alternative that both respects the rules of the game and ensures that all the opinions in the group will be considered: choosing C, an option that they all agree on, but taking the responsibility to defend option D as well during the whole-class discussion (turn 41). The others explicitly ratify such decision at turns 42-44, displaying their agreement both gesturally and verbally, Sabrina even verbalizing her action of putting the letter C at the same time as she does it.

In order to study the fifth indicator, the *dialogic strengthening of arguments*, it is necessary to look at these students' contributions to the class debate that follows this vote:

1. MO1   and why did you guys chose C instead of B/
2. PAM   we thought like C and B because: what we
3. SAB   °D\°
4. KEL   °D\°

5. PAM   oh yeah C and D because em: like we chose C because em:
6. LOU   ((laughs))
7. PAM   like (0.9) <((opening hands, turned to the sky)) oh i can't really explain> <((hands back to the table)) but like (0.5) however like like however like much time it's putting like (0.3) prod- like producing the water/ should be like (1.0) sold at a higher price like if it's like more better quality it should be sold at a higher price but if it's just (0.5) <((skeptical face)) regular water [i guess>&
8. LOU   [((laughs))
9. PAM   &like it should just be (0.7) like affordable\ and then we (0.2) thought D too because em: we thought that like less fortunate families shouldn't be like (0.5) punished not really punished but shouldn't like (.) have like a: <((moving hands)) lack of water> because (.) of like their jobs or whatever [their income
10. SAB   [because it's not always&
11. MO1   [their income/
12. PAM   [yeah\
13. SAB   &their fault that they're poor and then some people like inherit money/
14. MO1   hum hum\
15. SAB   so like (.) why should they get (0.4) the: like (2.1)
16. MO1   yeah\ [that's a good x\ and you guys had a different view/
17. SAB   [yeah\
18. T2   ((laugh))

This exchange shows a great *dialogic strengthening of arguments*. All students seem to feel responsible for the different ideas previously expressed at their table. These ideas have thus successfully traveled from individual contributions to group reasoning, and then ultimately into complex dialogic arguments expressed during the whole–class debate. Pamela initially makes a mistake, saying B (an option just discussed in the public debate) instead of D. The others help her by whispering the correct letter (turns 3 and 4). However, they also show great confidence in Pamela's ability to develop the group's ideas. Despite her many hesitations and long pauses, no one interrupts her. Moreover, Pamela is able to explain not only her own argument about the injustice of depriving people of water because of their poverty, but also those of others, such as the fact that producing better quality water involves more work. In terms of learning, she has not only acquired a new idea, but the whole rationale behind it. It is only when she seems to be finished that Sabrina allows herself to add something to what she is saying, with very little overlap. Sabrina does not repeat what Pamela has said but brings something really new, notably by elaborating on the merit principle, with a novel example, that of the heirs. The sentence "*it is not their fault*", together with the example of inheritance, constitute counter-arguments to the counter-argument mentioned by Louise during the group discussion ("*it's their fault that they are (.) poor| in a way*", turn 16). If we take up Toulmin's structural pattern of the argument (Toulmin, 1958), Pamela and Sabrina here co-construct a "*rebuttal*", an emblematic element of the most complex level of argument. Here, both girls have enriched their reasoning by anticipating a potential counter-speech, thanks to the previous group discussion. Finally, the collective laughter at their table in turn 18 appears to be the group's reaction to the tension produced by Sabrina's long pause, and indicates that all the four students feel relieved when the facilitator passes the conversational floor to another group.

3.1.4 *Cumulative talk* and disagreement avoidance

Through a second case study showing a typical case of *cumulative talk*, I will now specify the contours of this other type of group talk with regard to the 5 indicators defined above. When the participants do not engage into *exploratory talk*, the identity issues related to the communicative situation are more visible, notably because they constitute an obstacle to collective reasoning.

- *A typical case of* cumulative talk

This case involves the same four students as the previous one, in a discussion that took place earlier in the YouTalk, about the first opinion question, reproduced below (see fig. 7).

7) In your opinion, which potential sources of drinking water are the most promising for the future?

A The discovery of new fresh water deposits.

B Water that we don't use today (that which, is economized).

C Desalination of seawater.

D Climate change leading to more rain.

E New techniques for the depollution of water.

F None of these are promising : water is going to be in short supply and will become THE conflict of the 21st century.

Figure 7. Opinion question 1, which serves as a basis for Pamela, Sabrina, Kelly and Louise's cumulative dialogue.

Here is the transcript of their dialogue:

1. PAM i think it's
2. LOU B:\
3. PAM [((looking at Louise then at the screen))
4. KEL [((looking at the screen))
5. LOU cause i feel like people waste a lot of water
6. KEL yeah
7. LOU like washing [their dishes like&
8. PAM [yeah:\
9. LOU &before they put them in the dishwasher brushing your teeth [showering
10. KEL [or showers
11. SAB yeah\
12. LOU [showering
13. PAM [sho[wers yeah\
14. KEL [they take long (.) showers
15. SAB ((nodding head in the affirmative))
16. SAB <((nodding head in the affirmative)) uhuh>
17. PAM or like=
18. LOU =or just like other stuff
19. PAM people like when they brush their teeth (.) they leave the water running/ or like you wash your face whatever
20. SAB <((nodding head in the affirmative)) uhuh>
21. KEL ((nodding head in the affirmative))
22. LOU or like people that throw away like bottled water [or half the time like it's like it's not even finished and they'll just throw it away
23. KEL <((nodding head in the affirmative)) [yeah>
24. LOU [so i think it's [B
25. SAB [or they dump it out on the sidewalks
26. KEL or like washing your car
27. PAM yea[h:
28. LOU [oh yeah
29. SAB [((nodding head in the affirmative))
30. KEL [and then all the chemicals in it just go in the grass (.) <((turning her head)) which is not good
31. LOU [((laughs))
32. SAB [((nodding head in the affirmative))
33. T2 ((all the students move their body away from the center of the table))
34. SAB [((stretches))
35. LOU [((stretches))
36. LOU <((stretching)) so do we agree [on B/>

37. KEL   [<((getting   close   to   the
    stand)) so
38. PAM   <((showing the stand with her finger)) put B on the
    thing\>
39. SAB   ((puts B on the stand))
40. LOU   ((stretches))

Only option B is considered, proposed by Louise. There is no direct verbal acceptance neither opposition to it, as the other students simply nod their heads in agreement and provide a series of examples to illustrate it, which they develop throughout the dialogue. Moreover, even if they agree, the students show a genuine concern to ensure that everyone expresses their consent during the final decision-making process, starting at turn 36. Thus, indicators 1 (*justification of opinions*), 2 (*topical alignment*), and 4 (*cooperative decision-making*) are positive.

What makes this talk *cumulative* rather than *exploratory* is indeed the lack of *critical examination* of the expressed ideas (indicator 3). Students are not at all critical of what others are saying, even when some of the statements are off-topic or irrelevant, as at turn 25, moving on to a general discussion on everyday practices damaging the environment. Such typical characteristic of *cumulative talk* has already been highlighted by Lewis and Leach, who distinguished it from *exploratory talk* by the fact that students do not constructively criticize neither question each others' ideas, nor consider alternative viewpoints (2006, p. 1283).

What about the fifth indicator, the *dialogic strenghtening of arguments*? In the class discussion that follows this dialogue, only one contribution is made by a member of this group:

        SAB   because we waste a lot of water like in the shower and like washing your face

Sabrina, on this occasion, simply repeats her own example. She does not incorporate into her speech ideas previously formulated by other group members. Indicator 5 is clearly negative.

• *An* exploration *limited to the non-controversial side of the issue*
While these students are clearly not engaged into *exploratory talk*, they are indeed arguing. They are using a classic rhetorical process, based on induction, which consists in constructing arguments from examples (e.g. Amossy, 2006, p. 133-139). Their argumentative activity differs from that which takes place when a group develops *exploratory talk* in that it is limited to one dimension of the debate space, namely its non-controversial side. Thus, *cumulative talk* can be defined as a consensual exploration limited to one alternative of answer to the argumentative question raised. In this case, students focus on

common examples of environmental education that limit the discussion to a familiar *doxa* (Amossy, 2006, pp. 36–37), about which they readily agree. Unlike the first case presented, where the same students are engaged in *exploratory talk*, here they do not provide elaboration on previous examples, but generally just express their agreement by repeating it and adding another one (turns 10, 17, 18, 19, 22, 25).

Instead, they seem to be brainstorming to find examples that illustrate the chosen option. They accumulate them by frequently using the logical connective "or". Without context, their dialogue could be interpreted as answering a knowledge question, as they seem to consider the chosen option as "the right answer" with expressions like "it's B" in turns 1–2 and 24. Since this is the first opinion question, one possible explanation is that they have not yet fully understood the nature of the task. In any case, it would be simplistic to understand their attitude as reflecting a situation of agreement, because there are in the YouTalk corpus examples of groups reaching consensus and yet engaging in real *exploratory talk*. In such cases, even when they agreed on one option, students tended to consider the other options in order to develop counter-arguments and to get prepared for the class debate.

- *The perceived social relevancy of the type of group talk*

By comparing these two case studies involving the same group, it becomes clear that engaging in a type of group talk is not simply a matter of cognitive ability. If it were the case, students who do not show themselves capable of developing *exploratory talk* on the first opinion question would not magically become able to do so a few minutes later. But, if these students are already able to engage into *exploratory talk*, why do they deal with the first opinion question using *cumulative talk*? In fact, engaging in a form of group talk with high educational value should not only be understood as a cognitive-only process but also as a social practice, likely to be perceived as more or less relevant, in context, by the concerned students.

My interpretation of the difference between these two dialogues for the same group thus rather rely on a matter of contextual relevance. During the discussion of the first opinion question, students perceive the activity as consisting of demonstrating their knowledge of a consensual *doxa* typical of environmental education. They therefore adopt an identity position that insists on maintaining consensus within the group. However, when it comes to the third opinion question, they are willing to question each other's ideas for the cognitive advancement of the group. At this point, criticism is no longer perceived as a *face-threatening* act, as long as it strengthens the group's arguments in order to get prepared for the whole-class debate.

3.1.5 Hybrid case revealing the topical and sequential nature of group talk

This third case study on group reasoning is based on a dialogue between three French students: Jérémie, Julie and Laurent, about the main question of the YouTalk reproduced below (see fig. 8).

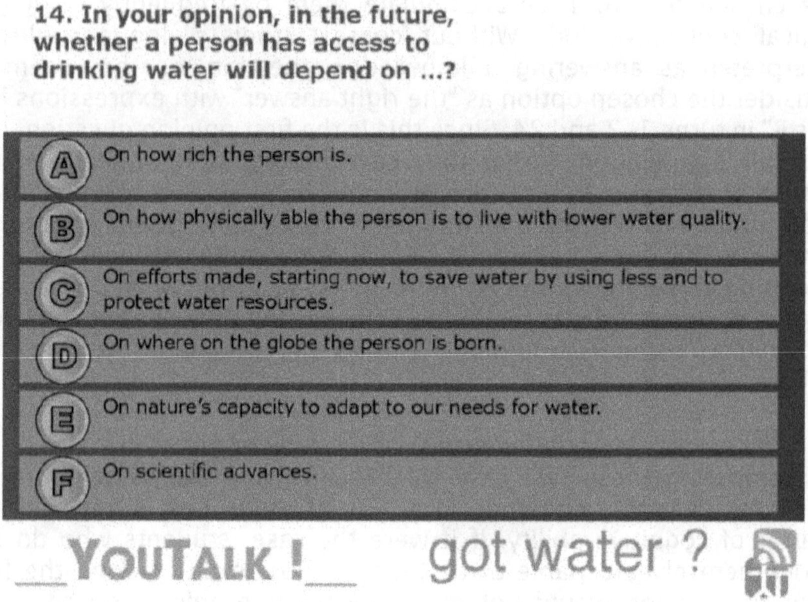

Figure 8. Main question, which serves as a basis for Laurent, Jérémie and Julie's hybrid dialogue.

The group discusses two options: A and F, through a talk that is 'hybrid' in several ways. At first glance, the students seem to alternate between *disputational* and *exploratory talk*. An analysis based on the first four indicators of group talk quality confirms such alternation. But paying a closer attention to the moments of *disputational talk* in this dialogue, I could specify what characterizes it. Finally, using the fifth indicator, the *dialogic strengthening of arguments*, and adopting a thinner analytical grain, I suggest a more in-depth analysis of the data, allowing for better understanding students' activity.

• *Arguing or quarelling: alternation between* dispute *and* exploration

The entire dialogue between the three students is transcribed and translated below.

112

1. JUL   euh: moi j'ai trouvé F xxx mais j'sais plus c'est quoi (euh: i found F xxx but i don't remember what it is)
2. JER   moi j'suis désolé mais ça va être A\   (me i'm sorry but it's going to be A\)
3. JUL   nan c'est [C\   (no it's [C\)
4. JER   [parce que: à l'heure qu'il est c'est pas rapport à A   ([because: at this time it is based on A)
5. LAU   ouais parce que l'eau elle va dev'nir d'plus en plus chère\   (yeah because water is gonna become more and more expensive\)
6. JER   et p- elle va dev'nir d'plus en plus chère et: les gens c'est des capitalistes et ça changera pas ça a toujours été comme ça et ça restera toujours comme ça\=   (and al- it's gonna become more and more expensive and: people are capitalists and it will not change it has always been like that and it will always remain like that\)
7. LAU   =l'eau ça a rien à voir avec le capitas-lisme\=   (=water it has nothing to do with capitalism=)
8. JER   =si\ parce que [c'est: les gens quand   (=yes\ because [it's people when)
9. LAU   [parce que l'eau c'est vital alors forcément ce c'est ça va dev'nir plus cher [même   ([because water is vital so obviously it it it's gonna become more expensive [even)
10. JER   [c'est vit- c'est vital   ([it's vit- it's vital)
11.   LAU [si c'était des communistes ou quoi ça s'rait pareil\   ([if they were communist or whatever it'd be the same \)
12.   JER   nan\   (it wouldn't\)
13.   LAU   si\   (it would\)
14.   JER   nan\   (it wouldn't\)
15.   LAU   l'eau elle deviendrait quand même chère\ (water would become expensive anyway\)
16. JER   nan\   (no\)
17. LAU   bah si tu veux faire comment/   (boh yes how do you wanna do/)
18.   JER   parce que   (because)
19. LAU   moins y'en a plus plus ça d'vient rare plus ça d'vient cher c'est logique\   (the less there is the rarer it becomes the more expensive it becomes it's logical\)
20. JER   bah moins y'en a plus ça d'vient plus ça d'vient cher (boh the less there is more it becomes expensive)
21. LAU   hein\   (hum\)
22. JER   ouais mais y'en aura toujours autant de: (yep but there will always be as much er:)
23. JUL   de l'eau   (water)
24. JER   de l'eau\=   (water\=)
25. JUL   =[mais après ouais mais ça=   ([but then yeah but it=)

113

26. LAU =[ouais mais après faut trouver des moyens justement pour euh: pour euh (=[yeah but then you must find ways precisely to uh: to uh)

27. JUL qu'elle soit propre/= (to make it clean/=)

28. LAU =récupérer d'l'eau d'la mer et tout ça (=use water from the sea and all that)

29. JER bah ouais et les moyens c'est quoi c'est la tunne (boh yeah and such ways are based on what on cash)

30. JUL nan\ c'est les avancées [scientifiques (no\ on scientific [advances)

31. LAU   [ouais c'est la tunne\   ([yeah it's cash\)

32. JER et pour avoir des avancées scientifiques on fait [comment/ (and to get scientific advances how do you [do/)

33. LAU   [mais nan y'a pas b'soin\= ([but no there's no need\=)

34. JUL =nan mais [vous voulez pas de mettre des billets sur l'bord d'l'eau et ça marche il faut des instr- avancées scientifiques (no but [it's not gonna work if you put bills on the waterfront scientific progress is needed)

35. JER   [il faut d'la tunne\ il faut d'la tunne donc [c'est la A: ([cash is needed\ cash is needed so [it's A:)

36. LAU   [nan mais on ils savent déjà faire hein
euh désaler l'eau désaliniser [l'eau\ ([no but we they already know how to do you know uuhh unsalt the water desalinate [the water\)

37. JER   [mais ça coûte cher\ ([but it's expensive\)

38. LAU oui mais après c'est vital alors tu t'en bats les couilles de [l'argent\ (yes but it's also vital so you
don't give a shit about [money\)

39. JER
[((pretending to count bills))

40. JUL   [les inventions scientifiques si tu les fais dans quelques années tu trouves un moyen pas cher euh de: ([if you do scientific inventions in a few years you find a cheap way to euh to:)

41. JER tu trouves un moyen/ vas-y trouves-le\ (you find a way/ go ahead find one\)

42. JUL nan mais j'suis pas [scientifique merci\ (no but i'm not a [scientist thanks anyway\)

43. LAU   [nan mais après après payer l'eau tu t'en bats les couilles de l'argent c'est vital t'as pas c'est pas un problème d'argent\ ([no but after paying for water you don't care about money it's vital you don't have it's not a problem of money\)

44. JER   <((prenant sa tête dans sa main gauche)) ou:i> mais: maint'nant   (<((taking his head in his left hand)) yea:h> but: now)

45. LAU   tu vas pas dire <((imitant)) oh bah non [c'est trop cher> (you're not gonna say <((imitating)) oh well no [it's too expensive>)

46. JER   [mais attends\ c — c'est c'est pas parce que j'suis pour [mais c'est   ([but wait\ i- it's it's not because i'm in favour of it [but it's)

47. JUL   [<((à *Jérémie*)) moi j'pense [que c'est les avancées scientifques\   ([<(((*to Jérémie*)) i do think [it's the scientific advances\)

48. JER   [j'dis c'est c'qui va s'passer\   ([i say this is what's gonna happen\)

49. LAU   hein/   (what/)

50. JER   j'dis que c'est c'qui va s'passer\   (i'm saying that's what's gonna happen)

51. LAU   de quoi/   (what/)

52. JER   c'est à cause de: de sa richesse\   (it's because of: of how rich you are\)

53. JUL   c'est: (it's:)

54. LAU   nan mais là c'est d'l'accès à l'eau potable hein   (no but here it's about access to drinking water you know)

55. JER   <((oui de la tête)) ouais\>   (<((nodding head in the affirmative)) yeah\>)

56. LAU   ça dépendra d'sa richesse/   (it will depend on one's wealth/)

(2.5)

57. JER   [ouais\   ([yeah\)

58. LAU   [ouais/   ([yeah/)

(2.7)

59. JUL   il suffit pas d'être riche\   (being rich is not sufficient\)

(2.2)

60. JER   bah bientôt si\   (boh soon it will\)

(4.7)

61. JER   <((à *Julie*)) t'sais que UN sur 6000 euh: de la population euh: du monde/>&   (<((*to Julie*)) you know that ONE out of 6000 uh: of the population uh: of the world/>&)

62. JUL   ((nodding head in the affirmative))

63. JER   &<((à Julie)) a 90 % d'la: d'la richesse mondiale\>   (&<((to Julie)) has 90% of the: of the world's wealth\>)

(1.6)

64. JUL   <((nodding head in the affirmative)) hum:\>

65. JER   tu sais c'que ça dire/ 'fin tu comprends/   (you know what it means/ do you understand/)

66. JUL   ((nodding head in the affirmative))

67. JER   <((oui de la tête)) et bah voilà\>   (<((nodding head in the affirmative)) well there you go\>)
68. JUL   et bah quoi/   (and so what/)
69. JER   et bah tu: <((non de la tête)) tu t'rends pas compte/> (and boh you: <((nodding head in the negative)) you don't realize/>)
(0.8)
70. JUL   nan mais ça à quoi à voir/   (no but what does it have to do with it/)
71. LAU   qu'est-ce qu'tu t'rends pas compte/   (what is it that we don't you realize/)
72. JER   nan c'est <((*et pose sa tête sur la table entre ses bras*)) pas grave vas-y nan>   (no it's <((*and puts his head in his arms on the table*)) it doesn't matter go ahead no>
73. LAU   qu'est-ce qu'i'y'a/ <((*en le secouant amicalement*)) qu'est-ce qu'i'y'a/>   (what's the matter/ <((*shaking it amicably*)) what's the matter/>)
74. JUL   mais EXPLIQUES-TOI c'est bon là on est en train d'parler [c'est un débat\   (but EXPLAIN YOURSELF it's okay, we're talking [it's a debate)
75. LAU
[ouais arrête de faire d'la merde là\   ([yeah stop messing around\)
76. JER   parc'qu'il y'a (.) sur 6000ème [de personnes   (because there is (.) out of 6000th [of people)
77. LAU   [<((riant))   6000ème>
([<((laughing)) 6000th>)
78. JER   de personnes=   (of people=)
79. JUL   =<((pointant son menton vers Jérémie)) oui mais ça ça fait quoi/>   (=<((pointing her chin at Jérémie)) yes but what does it has to do with it/>)
80. JER   bah ça fait que (.) <((posant sa main sur la table)) les richesses elles sont tellement inégalement réparties>   (it has to do that (.) <((putting his hand on the table)) wealth is are so unequally shared>)
81. JUL   <((incliant la tête vers le bas)) ouais\>   (<((bending her head down)) yeah\>)
82. JER   et qu'y'a des mecs qui veulent <((moulin main droite)) toujours s'enrichir s'enrichir [s'enrichir   (and that there are guys who want to <((right hand mill)) always get richer get richer [get richer>)
83. JUL
[ouais/   ([yeah/)
84. JER   <((fin du moulin)) s'enrichir> et <((lève les paumes vers le ciel)) jamais i:ls> <((geste de donner)) lâchent la tunne\>   (<((end of the mill)) and richer> and <((raises palms to the sky)) never do they:> <((gesture of giving)) give cash\>)

85. JUL   humm\ et bah quoi/ (humm\ and so what/)
86. JER   et bah j'dis <((hausse les épaules)) juste ça> pour euh:
pour dire ça\ (and well i'm saying <((shrugging his shoulders))
just this> to uh: to say it\)
(2.2)
87. JUL   mais ça a à voir avec 'fin ça:/ (but has is something to
do with well does it:/)
88. JER   <((haussant les épaules)) je sais ça a rien à voir mais j'le
dis quand même\> (<((shrugging his shoulders)) i know it has
nothing to do but i'm saying it anyway\>)
89. JUL   d'accord\ <((regardant l'écran)) oui et bah moi j'te dis
que:> c'est les sci — c'est les avancées scientifiques qui vont
faire qu'on aura la: euh: de l'eau\ (okay <((looking at the
screen)) yes and well I am telling you that:> it's the sci – it's the
scientific advances that are gonna have us get the: uh: water\)
90. JER   <((oui de la tête)) ben si tu veux\> mais: <((non de la
tête)) moi j'dis pas ça\> (<((nodding head in the affirmative))
well if you want to\> but: <((nodding head in the negative)) i
do not say that\)
91. JUL   pourquoi/ (why/)
(1.3)
92. JER   parce que\ pour des avancées scientifiques il faut d'la
TUNNE\ (because\ to get scientific advances you must have
CASH\)
93. JUL   oui mais il suffit pas d'avoir d'la tunne il faut euh: (yes
but having cash is not sufficient you must have uh:)
(1.0)
94. JER   hein/ (what/)
95. JUL   <((avance sa main vers Jérémie)) suffit pas d'avoir d'la
tunne\> <((main qui ponctue)) okay il faut des sous pour les
avancées scientifiques ça c'est sûr> <((main en position
« stop », paume vers lui)) ça j'suis d'accord> mais il suffit pas
d'avoir des sous j'veux dire tu vas pas <((le signifie dans
l'espace)) prendre un bac d'eau euh pollué> <((geste de poser))
mettre des sous à côté> faire <((imitant)) voilà:\> [t'vois faut
qu'y'ait des scientifiques qui qui vont dire bon ben on va faire
ça neineinein (<((advancing his hand towards Jérémie)) it's not
sufficient to have cash\> <((hand beats)) okay it's necessary to
have money for scientific advances that's for sure> <((hand in
"stop" position, palm towards him)) i do agree on that> but it's
not sufficient to have money i mean you're not gonna
<((spatially shaping it)) take a tank of uh polluted water>
<((gesture of putting down)) put money next to it> do
<((imitating)) that's it:\> [you see there must be scientists who
who will say well well we're gonna do that neineinein)

96. JER
[ouais (.) mais c'est les impôts: ça\   ([yeah (.) but there are
taxes: for this\

97. JUL   oui mais faut des scientifiques ce ça sert à rien d'être
riche si t'as pas d'scientifiques   (yes but you must have
scientists it it's useless to be rich if you don't have scientists)

(1.3)

98.   JUL   il faut d'la tunne ET des scientifiques mais faut avoir
des scientifiques aussi (cash AND
                    scientists are needed but scientists are also needed)

  99. JER   mais faut avoir d'la tunne (but cash is needed)

  100.   JUL [oui mais si t'as des scientifiques c'est euh
logique qu'il faut les payer ([yes but if you have
                    scientists it's euh logical they must be paid)

  101.   LAU [oui mais les scientifiques t'inquiètes pas ils
sont tranquilles ([yes but the scientists don't worry
                    they're relaxed)

  102.   JUL   ouais mais ils vont pas donner [d'leur poche
euh pour euh: (yeah but they won't pay out of their own
                    pockets euh to euh:)

  103.   LAU   [et euh ça va ouais\
nan mais les scientifiques j'pense ils
                    gagnent assez pour pas casser les couilles ([and euh
it's okay yeah\ no but the scientists i think they earn
                    enough not to break our butts)

  104.   JER   [c'est sû:r ([for su:re)

  105.   LAU [ils vont pas s'arrêter d'travailler oh bah merde
tant pis j'suis pas payé bah tant pis tout
                    l'monde meurt\ ([they're not gonna stop working oh
shit it's a shame i'm not paid bad luck everybody dies\)

  106.   JER   ((puts letter A on the stand))

  107.   MO1 à trois vous affichez un deux trois allez-y (on
three you put it up one two three go ahead)

  108.   LAU ((takes letter A off))

  109.   JER   ((takes it from him and puts it on the stand
again))

  110.   LAU °ouais il faut mettre ça\° (°yeah we have to put
this\°)

  111.   JER   °mais:° (°but:°)

  112.   LAU °fallait l'mettre maint'nant là° (°we have to put
it right now°)

  113.   JER   °tu veux mettre quoi toi/° (°what do YOU wanna
put/°)

  114.   LAU °A\° (°A\°)

115.JUL   °nan mais c'est pas 'rav\° (°no but it doesn't matter\°)

  116.   JER   °les scientifiques° (°the scientists°)

  117.   LAU NAN faut mettre A pelo: on était deux à l'dire\

118

(NO we have to put A dude two of us said so\)
118.JER on met les deux <(((posant F à côté de A)) on met les
    deux:> ([we put both <(((putting F next
to A)) we put both:>)
119.   LAU   mais barres-toi: (but get out:)
120.   ANI   eh s'il vous plaît là\ (eh please there\)
121.   JER   [(trying to put the two letters on the stand and
    facing trouble))
122.   JUL   [°nan mais: c'est bon° ([°nan but: it's okay°)
123.   JER   °nan mais arrête là:° (°no but stop there:°)
124.   JUL   °j'te jure ça va j'ai pas envie d'parler en plus\°
(°i promise it's okay and i don't wanna talk anyway\°)

Using the first four indicators of the quality of group talk, this
discussion can be described as hybrid as it alternates between
disputational and *exploratory* talk over 6 distinct episodes.
Three moments of *dispute* can be identified. First, the students seem
to be engaged into *disputational talk*, from turns 1 to 4: they state
opposing assertions, and repeat them without justifying them nor
adapting to what others are saying. Indicators 1, *justification of
assertions*, and 2, *topical alignment*, are therefore negative. The rival
option is rejected without exploring it and whitout making explicit any
real refutation. In turn 2, Jérémie shows no interest in knowing what
others think of option A, which he simply presents as a definitive
choice. Criteria 3, *critical examination of all ideas*, and 4, *cooperative
decision-making*, are thus also missing.
In turns 11 to 16, the dialogue between Jérémie and Laurent also
shows characteristics of *disputational talk*, even though they agree on
the choice of option A: the conflict is about the reasons behind this
choice.
Finally, turns 29 to 31 also show features of *dispute*, but this time over
the opposition between option A and option F.
However, this conversation also presents three episodes of true
exploration. From turns 4 to 10, Jérémie and Laurent are truly
collaborating. Laurent formulates two arguments justifying Jérémie's
assertion, which the latter immediately repeats: water will become
more and more expensive, and water is vital. Indicators 1
(*justifications*) and 2 (*topical alignment*) are positive. But Laurent, in
turn 6, simply rejects another reason given by Jeremy for choosing
option A, which results in a new episode of *dispute*. Thus, indicator 3
is partially positive: there is a critical look at the other's argument, but
no justification is given for the new statement proposed as a counter-
proposition.
It is only at turn 17 that Laurent finally engaged in exploring the reason
proposed by Jérémie to justify option A, by asking him to explain it
further, initiating a new episode of *exploratory talk*. In this second
moment of *exploration*, Laurent also develops one of the arguments

just mentioned earlier: "the rarer it comes the more expensive it is" (turn 19). This idea is criticized jointly by Julie and Jérémie, in a constructive way, who justify their view by rejecting the premise of water scarcity and arguing that there will always be the same amount of water on Earth (turns 22–24). Julie then reintroduces option F (access to water will depend on scientific advances) into the discussion, embracing the rule that a proposal cannot be abandoned without being truly examined, corresponding to indicator 3. Laurent's reaction demonstrates a concern for such rule, as he shows that he has not forgotten Julie's proposal, and even contributes to its reintroduction into the discussion. Indicators 1 (*justification of statements*), 2 (*topical alignment*) and 3 (*critical examination of all proposals*) are therefore positive here. On the other hand, Jérémie's behaviour, who then simply repeats the rival option, announces the beginning of another moment of *dispute*.

Finally, from turn 32 onwards, the students seem to engage for a longer time into a dynamic of *exploratory talk*. They fuel the discussion with interesting justifications that respond to each other. Jérémie uses an argument based on causality: he demonstrates that money is a primary cause, prior to scientific progress. Julie counters him with two rebuttals. The first one exploits the mechanism of reasoning by the absurd, and consists in caricaturing Jérémie's statements, simplified to the extreme, as the simple fact of putting money on the table, an action which cannot, on its own, make science progress (turn 34): this supposed first cause is thus presented as not sufficient. Julie's second counter-argument is based on the inversion of the causal direction proposed by Jérémie: it would be rather the progress of science which, progressively, would make a technology less expensive and therefore more easy to afford (turn 40). It is finally Laurent who manages to overcome such opposition on the material aspects of the issue, by appealing to the value of respect for life: "it's vital, so you don't give a damn about money", and by insisting repeatedly on the vital dimension of drinking water (turns 38 and 105). Thus, criteria 1, 2 and 3 are met, showing a group reasoning of quality. Indicator 4 is also positive: Jérémie shows great concern for taking into account the point of view of all the members of the group at the time of the final vote, including Julie, who defends the option that is rival to his own (turns 110, 113, 115). He is thus ready to put both options on the stand (turns 119 to 123). It is finally Julie herself who decides not to make her opinion visible to the rest of the class, a strategy to avoid speaking in front of the whole-class (turn 124).

- *Defining* disputational talk

A closer look at the three episodes of *dispute* identified in this discussion makes it possible to specify what distinguishes this type of group talk from *exploratory talk*, referring to indicators 1 to 4 of the quality of collective reasoning. In *disputational talk,* it is the repetition of opinions that takes precedence over their justification (indicator 1). This is particularly visible here in the first three turns of speech, when each student affirms the preferred option, without arguing his choice. This dynamic of repetition of assertions and counter-assertions also appears at turns 12 to 14 and 29 to 31. As a result, the second criterion, namely *topical alignment*, is limited: the students do not engage in *exploring* the ideas brought by others. They reject them globally, without *critically examining* the reasons behind them (criterion 3 is therefore also negative). Indicator 4 also seems to be "in the red": decision-making appears to be individual and offensive. Turns 2 and 3 are emblematic of this: the dialogue has just begun, but individual opinions are presented directly as irrevocable decisions, through the use of impersonal and indicative verbal forms. The students then act as if their opinions were obvious great truth. Now that the nature of *disputational talk* discourse has been clarified, Table 16 summarizes the characteristics of the 3 types of group talk according to the 5 indicators presented.

Table 16. Indicators of the quality of reasoning in exploratory, disputational, and cumulative talk.

	Exploratory talk	Disputational talk	Cumulative talk
1	Justification of opinions	Repetitions rather than justifications	Justification of opinions
2	Topical alignment	Limited	Topical alignment
3	Critical examination of propositions	Some rejected without exploration	Accepted without critical assessment
4	Cooperative decision-making	Individual decision-making	Cooperative decision-making
5	Dialogic strenghtening of arguments	Absent	Absent

The highly conflictive episodes described above as corresponding to *disputational talk* should predict an absence of dialogic strengthening of arguments (fifth indicator). However, taking this last indicator into account leads to a more refined analysis and provides a renewed interpretation that better explains students' cognitive activity in this dialogue.

• *The dialogic strenthening of arguments as a key indicator*
Let's look at the contributions of these students to the class debate that follows this group dialogue, to see how they have effectively made their own the ideas and arguments previously mobilized by others. This criterion of circulation of the content of the exchanges between the members of the group makes it possible to avoid a hasty judgment of their common activity which would be based only on an effect of

style, for example by systematically assimilating the use of a confrontational rhetorical style to *disputational talk*. In this case, two contributions to the class debate are made by members of this group.

The first one is initiated by the facilitator (MO1), and includes an intervention by the facilitator–observer (MO2). It is transcribed and translated below.

1. MO1 et là-bas vous aviez une réponse différente des deux autres c'était laquelle/ (and there you had a different answer from the two others which one was it/)
2. JUL euh c'était la D\ (euh it was D\)
3. JER °la D/ ° (°D/°)
4. JUL °ouais j'crois° (°yeah i think so°)
5. JER °nan c'était la F j'crois\° (°no i think it was F\°)
6. SYL t'avais la réponse D et euh donc euh pourquoi tu penses/ (you had answer D and euh so euh why do you think/)
7. JUL °ah attends\° (°ah wait\°)
8. JUL ((reads the question again))
9. MO2 pourquoi la D [alors/ (why D [then/)
10. LAU °c'tait la F [t'avais dit° (°it was F [you've said°)
11. JER [°des avancées scientifiques\° ([scientific progress\°)
12. JUL °ouais c'était la F\° ouais c'était la F\ (°yeah it was F\° yeah it was F\)
13. MO1 comment/ (what/)
14. JUL la F\ (F\)
15. MO1 t'avais choisi la F/ (you chose F/)
16. JUL ouais\ (yeah\)
17. MO1 d'accord bah tu peux nous expliquer pourquoi la F/ parce que justement elles étaient plutôt contre la F elles pensaient que ça dépen- la F euh rev'nait aux rich[esses euh (okay so can you explain why (you chose) F/ because they were specifically against F they thought it would depen- F equals wealth euh)
18. JUL
[ben la F euh c'est sûr qu'il faut d'l'argent mais euh on peut pas 'fin c'est pas en ayant juste de l'argent que on: qu'on va avoir d'l'eau: potable quoi 'fin pour avoir assez d'l'eau faut aussi euh ([well F euh for sure money is needed but euh we cannot i mean it is not just having money that we: we're gonna have water: drinking water you know i mean to have water we also need euh)
19. JER des scientifiques pour euh la faire (scientists to euh make it)
20. JUL ouais (yeah)

21. JER   'fin pour trouver des avancées: des trucs comme ça (i mean to make progress: such things)

Julie is asked to justify her answer. She seems rather confused about having to speak in front of everyone: she can't even remember the letter that corresponds to the option that she was defending, she actually had insisted on not displaying it precisely to avoid speaking in the class debate. Jérémie and Laurent help her and show that they know which option she had chosen (turns 5, 10, 11). Jérémie, in turn 19, even complete Julie's sentence, paraphrasing the counter-argument she had specifically used in opposition to his own. Moreover, Julie's contribution integrates the criticisms opposed to her opinion during the group discussion. She repeats a dialogic argument then constructed (turn 18), which concedes that money is necessary but not sufficient for the development of scientific progress. Such a concession constitutes a rebuttal in the sense of Toulmin's structural pattern of argument (Toulmin, 1958,), i.e. a counter-evidence anticipating an opposing discourse. As in the typical case of *exploratory talk* presented above, we find here an argument previously shared in group, which has been strengthenned by the integration and anticipation of a potential counter-discourse (indicator 5 is positive).

The second contribution of this group to the class debate that follows the studied dialogue is shorter, and entirely made by Laurent:

LAU   moi j'pense que l'accès à l'eau elle va dépendre de la richesse des gens parce que plus elle va dev'nir chère 'fin plus elle va dev'nir rare plus elle va dev'nir chère et donc euh: ça s'ra les riches qui: auront (i do think that access to water will depend on people's wealth because the more expensive it's gonna become well the rarer it's gonna become the more expensive it's gonna become and so eh: it will be the rich peopl who: will have)

Here, Laurent only reports his own idea, without taking into account any counter-argument, even though such idea was opposed a rebuttal during the group discussion (turns 20-25). Thus, this contribution does not meet the fifth criterion of a quality group talk. This indicator shows that when discussing the possible reasons for choosing option A, the students were not actually cooperating in an effective and constructive way, so they were not really engaged into *exploratory talk* in turns 18-29, even though they were already no longer in a typically *disputational talk*.

The circulation of arguments from one member to another, and from the small group to the class group, and their associated *dialogic strengthenning*, thus appears to be a decisive criterion for assessing the quality of group reasoning. In the present case, the two contributions of the group to the class debate differ greatly, revealing a variation in the quality of collective reasoning about two different topics.

The first contribution proves that, when the students were debating about the choice between option A and option F, they managed to build a common reasoning and to appropriate the ideas brought by each member of the group. In this way, we can revise the analysis and consider that all the turns of speech concerning this topic deserve to be interpreted as a global, coherent sequence of *exploratory talk*. The two episodes of *dispute* between A and F, appearing at the very beginning of the discussion, and during the reintroduction of the subject (turns 1–2 and 30–32) can therefore be understood as opening and reopening subsequences serving to exhibit the rival points of view before their collective *exploration*.

In contrast, Laurent's contribution to the class debate that follows this group dialogue, about another topic, namely the reasons for choosing option A, shows that in turns 3–29, students were not really *exploring* this issue together. This part of the discussion can be reinterpreted as an overall topical sequence of *dispute*, in which two more collaborative local subsequences are embedded. The first one (turns 4–10) should be understood as an opening, rather *exploratory* subsequence, which allows Laurent and Jérémie to identify their points of agreement and disagreement before engaging into *disputational talk*. The second one (turns 18–29) can be characterized as a transitional subsequence, in which each participant gradually goes back to a behavior that makes possible the *exploratory talk* sequence that follows. In particular, it is during this transition that Julie steps back into the discussion, and reintroduces option F, producing a topical shift towards the opposition between options A and F.

Finally, this fifth indicator, *the dialogic strengthening of arguments*, allows us to understand deeper what is happening in this 'hybrid' dialogue, by distinguishing between stylistic effects and real argumentative mechanisms. The new interpretation of the data points to both the sequential and topical nature of the group talk and associated collective reasoning. Figures 9 and 10 summarize this final interpretation of the dialogue between Jérémie, Julie and Laurent. From a methodological point of view, this case also reveals the interest of distinguishing three scales of analysis for an in–depth understanding of group argumentation: that of the dialogue, corresponding to the whole discussion on a given issue; that of the topical sequence, encompassing all the parts of the dialogue dealing with one of the subjects addressed during this discussion; and that of the subsequence, corresponding to the parts of a sequence that can locally fall under another type of talk, in order to serve specific local interactional functions (opening, transition, closing). Thus, this third dialogue turns out to be hybrid in three ways: 1) it is mainly composed

124

of two discursive sequences, one of *exploratory talk*, and the other of *disputational talk*, dealing with different topics; 2) some subsequences locally present characteristics of a type of talk that differs from the nature of the talk for the global sequence to which they belong; 3) it presents a subsequence itself of a hybrid nature, interpreted as a transition between the two global topical sequences.

Exploratory sequence about the A/F debate.	
1-3	Disputational > Expository pre-sequence about the competing options (A vs F)
4-11	Exploratory > Expository pre-sequence about what they disagree on about resaons for A
11-16	Disputational, about reasons for A
17-28	Embedded transitional sub-sequence, gradual alignment to constructively critical footing
29-31	Disputational > Expository pre-sequence of still competing options (A vs F)
32-60	Exploratory, about A vs F
61-90	Embedded disputational sequence, about reasons for A
91-110	Exploratory, about A vs F
110-146	Closure, with gradual disalignment of individual footings

Figure 9. Topical sequence of exploratory talk about the opposition between options A and F.

Disputational sequence about reasons for A.	
1-3	Disputational > Expository pre-sequence about the competing options (A vs F)
4-11	Exploratory > Expository pre-sequence about what they disagree on about resaons for A
11-16	Disputational, about reasons for A
17-28	Embedbed transitional sub-sequence, gradual alignment to constructively critical footing
29-31	Disputational > Expository pre-sequence of still competing options (A vs F)
32-60	Exploratory, about A vs F
61-90	Embedded disputational sequence, about reasons for A
91-110	Exploratory, about A vs F
110-146	Closure, with gradual disalignment of individual footings

Figure 10. Topical sequence of disputational talk between Laurent and Jérémie about the reasons for choosing option A.

3.1.6 A grid for the Multi-level Analysis of Group Talk (MAGTa)

These three case studies show that the 5 indicators defined above are effective in specifying the nature of each type of group talk, either *exploratory*, *cumulative*, or *disputational*. They allow not only to characterize typical dialogues that fall unequivocally within one of these types of talk, but also sequences or subsequences of *hybrid* dialogues, thus helping to better understand what is going on at the interactional and cognitive levels. This is only possible if we consider reasoning as a phenomenon based on the interaction of three social levels: individual contribution, group dialogue, and whole-class debate. Such results inspire a grid for the Multi-level Analysis of Group Talk (MAGTa), a conceptual and methodological toolkit allowing, beyond this specific case study, to apprehend the dynamics of collective reasoning in any small working group. The challenge, is to provide avenues of reflection to understand why a group tends to engage in a certain type of discourse, at a given time, in a given context.

Above all, the prevalence of *hybrid* dialogues over *typical* dialogues (43/76 or 56.6% versus 33/76 or 43.4% cf. Table 13) shows that it is far from easy for students to form a real group engaging collectively in one type of talk. Even when students manage to form a group and maintain the same type of group talk throughout a dialogue, they do not necessarily engage into the one with highest educational value, namely *exploratory talk*. Contrasting the first two dialogues, involving the same group of students, shows that engagement in a type of talk is not only a matter of cognitive ability but also depends on the perception of a social context as expecting a particular type of talk. The identification of a hybrid group talk subsequence, playing the role of transition from *dispute* to *exploration*, in the third case, shows, moreover, that collective engagement into a type of talk can take time, and requires discursive and interactional work for group members to align their individual behavior. Such results lead to two methodological concerns: the need to multiply the scales of analysis of group talk, and the fact that groups, and the quality of their talk, should not be considered as fixed entities, but as dynamic constructs resulting from inceasing active adjustment between the their members.

- *Multiplying analytical scales*

Multiplication of the analytical scales is a raising issue in research on dialogic education. For example, Hennessy and her colleagues distinguish between the 'communicative situation' the 'communicative event' and the 'communicative act' in their approach to educational dialogue, although the codes that they have developed are systematically applied only at the level of *communicative acts* (Hennessy et al., 2016, pp. 3-4). As mentioned earlier, I prefer to use the categories of "dialogue," "sequence," and "subsequence" for two reasons: 1) they are more transparent, closer to common language; 2) they better refer to the the structure of discourse and the hierarchical relationships existing, for example, between sequences and subsequences.

The category of *dialogue* as I use it here corresponds to an intermediate temporal unit between what these authors call 'communicative situation' (which corresponds to a pedagogical activity, and would here come back to YouTalk as a whole) and 'communicative event'. A *dialogue* must be understood, in the framework of MAGTa, as a group interaction, beginning with specific instruction explicitly inviting to a discussion–based task, and ending with this task as the person in charge of the didactic framework (here the moderator) move on to another task.

The term *sequence* refers to a topical unit within a dialogue. An entire dialogue may consist of a single sequence, as in *typical* dialogues, or several sequences of different types of group talk, as in *hybrid* dialogues. Although it is another analytical scale, the *sequence*, like the 'communicative event', is defined by its topicality.

Subsequences are parts of sequences that play specific interactional functions, and thus may have characteristics that differ from the overall sequence to which they belong. *Subsequences*, like 'communicative acts', are defined by their interactional function, but they correspond to larger units, usually comprising several speech turns, whereas the latter consist of only one turn or part of a turn.

- *Relation between individual communicative behaviour and group talk*

The group talk results from the communicative attitudes of its individual members, attitudes that it reciprocally shapes somehow. I call the process that relates these entities the *alignment* of individual footings. The interpretative hypothesis underlying this conceptual tool corresponds to the idea that the problem of the preservation of the face of the individuals strongly structures their perception of one or another communicative attitude as relevant, and thus their *alignment* on a certain type of group talk.

In the context of small groups, students must choose the appropriate politeness system between two distinct options: the politeness corresponding to ordinary conversation between peers, or that corresponding to an argumentative situation, as needed for the task. Depending on whether they perceive their activity as belonging to one or the other of these politeness systems, they will set up specific ways of preserving their face, corresponding to an *identity footing*[12] , which determines their communicative attitude, and consequently the type of talk into which the group engages. When all group members are aligned on the same *identity footing*, demonstrating the associated communicative attitude, they manage to engage into a typical group talk. Table 17 presents how a particular identity footing corresponds to a specific communicative attitude, and the associated type of group talk.

---

[12]The term 'footing' here is used in the sense of Goffman (1981): it is a role assumed by an individual during a conversation, emphasizing some aspects of his social, institutional or interactional identity. An individual may change such footing several times during a given interaction.

In *cumulative talk*, as in the second case studied, the preservation of *face* is ensured by a communicative attitude consisting in avoiding the expression of disagreement, which corresponds to a *consensual* identity footing, typical to ordinary conversation. Conversely, in *disputational talk*, each student strongly relates his or her *face* to his or her own ideas and their victory over others', which corresponds to a *competitive* identity footing. What makes *exploratory talk* so special is that the problem of face preservation is transferred to the group level, with each member relating his or her *face* to the cognitive advancement of the group. As in the first case studied, individual criticisms and challenges appear only insofar as they allow the construction of stronger arguments, considered to be the property of the whole group, and sources of pride for each of its members. When the need for social recognition is satisfied this way, everyone is on a *constructive critical* identity footing. The concern for one's own face being transfered to the group, each one can focus on something else, namely on the cognitive dimension of the interaction. The emotions displayed regarding face preservation indeed play a crucial role in mediating such interactional alignment on more or less constructive group sociocognitive practices (Polo et al. 2016b, Polo et al., 2017a, see 2.2).

Table 17. Individual identity footing and face preservation system associated with each type of *group talk*.

	Type of group talk		
	Cumulative talk	Exploratory talk	Disputational talk
Identity position	consensual	constructive critical	competitive
Face preservation system	preservation of consensus by not expressing disagreement	focus on group success	search for the victory of one's own ideas over others'

Such a theoretical assumption helps understand the existence of fully *hybrid* subsequences, in which students display different *identity footings*, their lack of *alignment* not allowing for engagement into a typical group talk. For example, in the transition subsequence identified in the third case study, Jérémie, Julie, and Laurent are in the process of aligning themselves on a *constructive critical* identity footing, necessary to *exploratory* talk. The students are gradually adapting their communicative attitude. They accept to consider the proposals of others (turns 18 and 23), but they are not ready to submit their own assertions to the collective examination yet. For example, Jérémie does not explain why he disapproves of capitalism or how this statement relates to option A (*access to water will depend on wealth*)

128

(turn 19). Laurent does not listen to the refutation of his idea that water resources are decreasing, as evidenced by his later contribution to the whole-class debate. Julie is just coming back into the discussion and is not really giving her point of view yet. Although the students are not yet ready to have their ideas criticized and challenged, they are already open and willing to consider those of others, which is an important step in gradually aligning their identity footing on a *constructive critical* attitude. Finally, from turn 29 onwards, the students make their reasoning more explicit and accept to submit it to others' criticism. Despite keeping a relatively confrontational rhetorical style, their identity footing is primarily focused on collective cognitive advancement, even if this means abandoning or modifying initial individual ideas (e.g., turns 32, 36, 98, 100).

- *Implications for instructional design of activities fostering collective reasoning*

Such understanding of the sociocognitive dynamics of groups dealing with argumentative tasks make it possible to set up pedagogical activities that encourage them to engage into a genuine cognitive *exploration* of a problem. Thus, it is essential that any argumentative activity with an educational aim establishes a social context fostering *exploratory talk*, making its use obviously relevant: a place where any idea can be formulated, and where rival options are examined without threatening the faces of the people carrying them.

A first approach to this can rely on a perspective of explicit teaching. In the manner of the *ground rules* mentioned by Mercer and his colleagues, one can imagine that the person in charge of the didactic framework (in this case, the facilitators, but it could be a teacher) carry out a preliminary work, preceding the activity, setting up collective rules aiming at keepend students' faces safe in order to foster *exploratory talk*. These rules could be regularly reminded throughout the exercise. They would mainly involve encouraging information sharing, participation of all group members, respectful individual challenging, interest in others' ideas, and a concern for evaluating arguments and not individuals (Asterhan, 2013; Fernández et al. 2002; Mercer & Sams, 2006; Wegerif et al., 2004).

Another way to encourage *exploratory talk* might be to constrain the form of group dialogue through the use of a specific '*script*' that forces compliance with these rules. For example, the group could be provided with a specific procedure involving the successive discussion of options A to F, with each student having to decide on each option before moving on to the next (Weinberger, et al., 2005). Such '*scripting*' is not necessarily chronological: group members may be given speaking tickets to encourage everyone's participation, or only the decision-process may be scripted to ensure cooperation in the process.

Still, one should remember that such tools should not simply banish any possibility of *disputational* or *cumulative talk*, which may play specific local role in a global exploratory sequence as shown in the third case study. Similarly, it is important to respect the time that group members may need to align their communicative behaviour and underlying identity footing towards *exploratory talk*, trhough hybrid transitions. Thus, providing systematic negative feedback on non-optimal attitude at the beginning of the activity may destabilize the students and disturb the normal process of opening an *exploratory* sequence.

Inviting a group to discuss a controversy is not sufficient for a real collective reasoning to occur. In the best scenario, the group engages into a cognitive *exploration* of the question, developping a talk that makes explicit, considers and evaluates any idea imagined by its members, elaborating on the most promising ones.

However, the discussion sometimes rather turn to *cumulative* or *disputational* talk. In the first case, only the non-controversial side of the question is discussed: consensus seems as the most relevant way to ensure *face* preservation. In such a configuration, there is no room for critical thinking. On the opposite, in *disputational talk*, nothing is done to preserve one another: without really considering the alternative propositions, each person focuses on tough defense of his or her own opinion. Preserving the *face* then means imposing his or her ideas over others'. At the end of the day, engaging into *exploratory talk* implies another regulation of the matter of social recognition: by creating a safe climate of discussion, all the members of the group aligning on a *critical constructive* self-identity footing. *Facework* is then dissociated from individual ideas and instead linked to group cognitive advancement. Such an alignment process might take long, which explains why hybrid subsequences of group talk, not falling into any of these three categories, might occur. Finally, several case studies show that engaging as a group into one type of talk or another is not determined by individuals' cognitive capacity but rather depends on the perceived social relevancy of talk in a given context. These results should be taken into account for designing and implementing pedagogical tools for collective reasoning: 1) from the beginning, by presenting the communicative situation in a way that would make critical constructive attitude perceived as the relevant self-identity footing to display; 2) at giving instructions, in order to make explicit the ground rules for *exploratory talk*; 3) and even possibly before the session, by scripting high-quality argumentation action patterns within the activity itself (for instance: any contribution should explicitly mention (dis)agreement with previous one, or be justified, etc.).

3.2   Social and cognitive functions of group emotions

Even if it can be useful for analysis, at some point, to separate the affective from the social dimensions of group argumentation, in order to focus on one or the other, these two aspects are strongly intertwined in the studied interactions. As above-mentioned, the social tends to "invade" the affective and *vice versa*. Indeed, the emotional framing of the problem results from a collective elaboration throughout the alternance of speech turns, and reciprocally constitutes *ad hoc* groups around a thesis, each one defending one of the rival opinions. Therefore, affective framing cannot be perfectly independent from the relational and social dynamics among the students. In the same way, the quality of a group's talk, and its propensity to engage into cognitive exploration, relates to how its members develop and display a specific identity footing, determining the emotions that are likely to emerge in the interaction. Engaging as a group into quality reasoning implies that each member adopts a constructive critical attitude. Such alignment into a critical constructive footing activates a system of social recognition in which there is no shame in formulating still ill-structured ideas, nor in changing one's mind, nor in being aggressive in criticizing others' points of view, nor even in being sad about failing to convince others of our own initial intuitions. On the contrary, it should promote the satisfaction of abandoning one's initial ideas for other, stronger, collectively constructed ones, which corresponds to an intrinsic motivation. I present here, after a quick review of the literature, a first model to describe the links between these different aspects of collective argumentation, in which emotions play a central role in mediating the group's socio-cognitive activity. I then illustrate how this theoretical tool can be operationalized by applying it to examples taken from the YouTalk corpus.

3.2.1 Research on emotions in argumentation

In recent decades, research on learning has undergone two major shifts, as a similar tendency questioned the field of argumentation theory. The first major shift, the 'social turn', consists in the extension of the concept of cognition from an individual perspective to a collective, even sociocultural perspective, notably with the emergence of "group *cognition*" (Stahl, 2006), as the paradigm of pragma-dialectics raised in argumentation theory (Eemeren & Grootendorst, 2004). The second, 'affective turn' has allowed the integration of the emotional dimension of learning into the agenda of educational research, with a renewed view of cognition reconciling reason and emotion. Some argumentation theorists have also made a similar shift, or at least asserted the need to include emotions in argumentation models (e.g., Plantin, 2011).

- *Emotions in empirical approaches to argumentation studies*
Historically, the institutionalization of argumentation studies as a field of research, at the end of the xxe century, conveyed a critical normative look at emotions, mostly considered for their fallacious potential (e.g.

Hamblin, 1970, Walton, 1992). Only quite recent empirical approaches, rehabilitating some aspects of ancient rhetoric, started to study emotions as argumentative resources (e.g. Micheli, 2010; Plantin, 2011, 2015; Hekmat, Micheli, Rabatel, 2013). Their descriptive goal is to understand the role of emotions in authentic argumentative discourse (Plantin, 2011, 2015), either studying people's felt emotions or the emotional tonalities conferred to discourse objects through *emotional framing* (Polo et al. , 2016, Polo et al. , 2013b, 2017a, 2017b, cf. 2.4.1).

- *Emotions in research about collaborative learning*
In the research field of collaborative learning, it is generally accepted that emotions play an important role in the socio-cognitive processes that enable learning. This literature mainly recognizes two impacts of emotions related to argumentation on collaborative learning. On the one hand, emotions are discussed for their positive impact in generating socio-cognitive conflict (Roschelle & Teasley, 1995, Doise & Mugny, 1981). These beneficial effects have been studied for conceptual or practical change, deepening of the debate space, or knowledge building (e.g. Andriessen, Pardijs, Baker, 2013, Sins & Karlgren, 2013). Conversely, some studies instead emphasize the detrimental effects of some emotions emerging in argumentation on group performance. The explicit disagreement necessary for socio-cognitive conflict, constituting a non-preferred act in conversation (Pomerantz, 1984), is a source of tension that can be difficult for the group to manage. In this context, students may be led to develop relaxation strategies that do not promote learning (e.g., Andriessen, Pardijs, Baker, 2013).

These ambivalent results about the impact of argumentation-related emotions have led scientists and educators to address the questions of real-time *emotion* awareness and *emotion* regulation (e.g. Järvenoja & Järvelä, 2013). The research objects called emotions here correspond to the feelings of the people involved in a learning situation. Nevertheless, most studies rely primarily on discursive indicators. Two main oppositions structure the study of emotions in collaborative learning: the focus on individual feelings (shame, motivation, etc.) *vs.* on emotional events occurring at the group level (confidence, feeling of efficiency, etc.); and the distinction between long-term emotional constructs (collaborative climate and group history) or more punctual ones (emotion aroused by a specific task or subtask).

Table 18 shows the main characteristics of these two research traditions about argumentation-related emotions.

Table 18. Emotions in argumentation studies and collaborative learning.

	Argumentative studies		Collaborative learning	
Focus	Fallacious strategies vs argumentative resources		Effects on reasoning Individual level & group level	
Method	Discourse analysis		(Mainly) Discourse analysis	
Object	Expressed emotions		Felt emotions	
	people's feeling	objects' tonality	long-term	local constructs

3.2.2 The role of emotions in collective reasoning: a model

Figure 11 presents a model that articulates these theoretical inspirations in order to propose an overall view of the role of emotions in multiparty reasoning. Their social functions (in black) are distinguished from their cognitive functions (in white), although the two types of functions do jointly work during a given socio-cognitive activity. An interaction never completely starts from scratch. Each person arrives with his or her own internal emotional state, which includes pre-existing feelings about the objects under discussion and the people involved in the task. At the cognitive level, the initial formulation of the question at stake also constitutes to an *a priori* framing of the activity, which is not emotionally neutral. When the interaction starts, some aspects of these pre-existing elements are selected, and given a communicative form to be shared between the participants. Then, two concomitant phenomena take place, one social, the other cognitive, giving rise to two forms of emotional speech: the expression of the feelings of the present participants and the emotional framing of the discourse objects. These two affective processes evolve continuously during the debate, in real time, each participant adjusting his or her communicative behaviour to the emotions expressed by others and specifying his or her view on the problem through a specific *emotional schematization*. These two phenomena refer respectively to the social and cognitive functions of emotions in group argumentation.

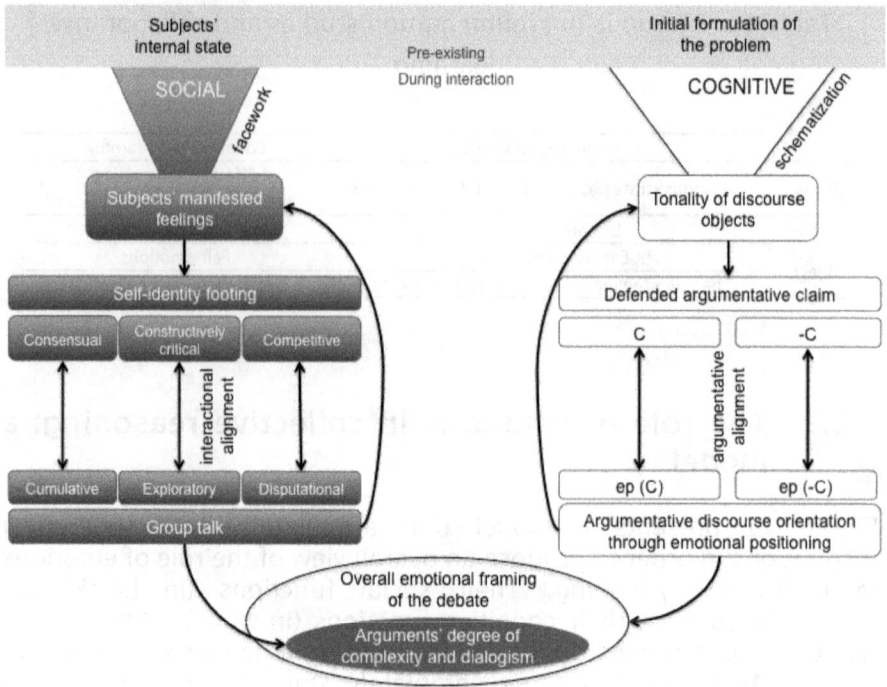

Figure 11. The functions of emotional entities in the social–cognitive activity of collective reasoning.

- *Emotional entities at the individual level*

On the social plane, the ongoing system of face preservation and social recognition is embodied in members' endorsement of a given identity footing, which constrains the possibilities of emotional expression. Students manifest and interpret their feelings and those of others differently depending on whether they assume a *consensual*, *competitive*, or *constructively critical identity footing*.

On the cognitive plane, each person defines and categorizes the problem during the discussion by emphasizing some aspects rather than others, according to a process of *schematization* that implies an emotional dimension. Thus, the objects discussed, when they emerge in the discourse, are given an emotional tonality that is argumentatively oriented towards the defense of a thesis (Polo et al. , 2013b, 2017a, see 2.4). In the corpus studied, this thesis is one of the competing response options (A, B, C, D, E or F). For the sake of simplification, and to highlight the rivalry between the defended theses, only two possibilities appear in the model: C and –C.

- *Resulting sociocognitive alignments at the group level*

In a polylogal argumentative activity, each person adjusts his or her individual emotional position (identity footing on the social plane and emotional framing on the affective plane) through processes of alignment or misalignment. *Interactional alignment* allows participants to engage, as a group, into a specific type of talk: *disputational talk*, *cumulative talk* or *exploratory talk*. When all group members are aligned on a *consensual* identity footing, the group accumulates ideas; when they are aligned on a *constructive critical* footing, then it truly explores the issue, and when they are aligned on a *competitive* footing, then the group faces a dispute.

Argumentative alignment occurs when people defending the same thesis develop similar emotional position about the issue under discussion, *schematizing* it in a similar way. An emotional position associated with the defense of option C emerges, ep(C). Its counterpart, for those defending the rival option –C, is the emotional position ep(– C). The group at play here is not necessarily the one physically gathered around the same table, it is rather an *ad hoc* entity based on intellectual affinity, which may involve students from different table groups, though sharing the same opinion. Argumentative rival sides emerge in the classroom, opposing students defending alternative theses and consistent emotional positions. Each contribution by a supporter of C (or –C) argumentatively steers the interaction towards that thesis by provoking a *phasic* emotional move towards ep(C) (or ep(– C)).

- *Class–level outcomes*

Finally, these *phasic* moves throughout the alternance of speech turns end up drawing a background emotional tonality to the debate, or *thymical* framing, which may differ from the emotional framing initially suggested by the activity, on the cognitive plane. On the social plane, the type of group talk that prevails has an impact on the quality of the arguments constructed, on their degree of complexity and dialogism. When the group engages into *exploratory talk*, students strengthen and complexify their reasoning by incorporating others' counter-arguments.

3.2.5 Relationships between the social and the cognitive functions of group emotions

If the emotions in group argumentation play social and cognitive functions through the above–described mechanisms, these two dimensions are not perfectly independent from each other. It is important that the model how emotions work in group argumentation accounts for the such interactions these social and cognitive functions. The easiest way to highlight this problem and to generate related working hypotheses is to go back to authentic data. Below are two case studies that suggest the existence of a relationship between the *thymical* tonality of the debate (cognitive plane) and the nature of group talk (social plane).

- *Case study:* cumulative talk *and low emotional framing*
The dialogue between Louise, Pamela, Sabrina and Kelly discussing OQ1, presented above as emblematic of cumulative talk (cf. 3.1.4), is characterized, in terms of *thymical* emotional framing, by a very low intensity. Later, when these students talk about OQ3, and engage into exploratory talkl, the *thymical* intensity of their exchanges is much higher. This comparison suggests that a minimal *thymical* intensity is required for individuals to move beyond ordinary conversations focused on maintaining social link, and engage into verbal interactions focused on cognitive objects, especially if these objects are controversial.

Thus, OQ1 is treated by the four girls in a very academic way, with limited personal investment. They ignore any potentially controversial dimension of the question, and appeal to the common *doxa*, well known to all, of environmental education, by listing the "good practices", even if it brings them away from the topic of drinking water. The students then show clear signs of disinvestment: they take long breaks, look at their watches, stretch or yawn (cf. 3.1.4, turns 14, 19, 30, 33–36, 40).

Cognitively, students develop a very low emotional framing of OQ1, which is particularly visible through the parameter of the distance from the concerned people. The girls present the problem of drinking water, throughout this dialogue, as relatively distant from them. They do not involve themselves directly in the discussion, and mainly use the third person to talk about "*people*" who might waste water (cf. 3.1.4, turns 5, 7, 9, 14, 19, 22, 25). The second person "*you*" is used only twice, in a general sense (cf. 3.1.4, turns 19 and 26). The more involving pronoun "*we*" only appears during group members' participation to the whole class discussion that follows this dialogue, also in a general sense. Finally, the third person plural "they" refers, in their discourse, only to the members of a collective entity that wastes water, who should strive to limit such waste. Finally, when it comes to OQ1, an emblematic dialogue of *cumulative talk*, these students never present themselves as potential victims of the lack of drinking water.

Conversely, when these students address OQ3, via a true *exploratory talk*, they construct in their dialogue a much shorter emotional distance to the problem, thus producing a more intense *thymical* framing, which "warms up" the topic. The problem of determining the price of drinking water is presented as an issue that directly concerns them, as it does everyone (cf. 3.1.3, "all people", turn 14, "everyone", turn 18). For their general formulations, they then prefer the second person, more involving, than the third (cf. 3.1.3, turns 1 and 8). An important part of the debate concerns whether access to water should be unconditional, especially from the perspective of not excluding the poor. At this point, Louise is the only one to keep to the third person, an emotional position, which is consistent with her opposition to the proposition that the price of water should depend on family income. This movement of *phasic* distancing here clearly reflects the argumentative purpose of her speech. On the contrary, Sabrina uses the second person in turn 8: "you *need water*" (tu as/on a besoin d'eau). In doing so, she frames the affordability of drinking water not as the problem of a few people distant from the participants in the discussion, but as a global concern that affects everyone, including themselves. Sabrina's sentence also tends to "warm up" the discussion by referring to the idea of necessity. She *frames* the issue not just as a moral dilemma or an ecological principle, but as a concrete problem of material survival, which pulls the emotional framing toward the "death" pole on the valency axis. Similarly, Pamela, at turn 9, aligning herself with the *phasic* rise produced by Sabrina, presents the risk of running out of water in a radical formula, as the total absence of water (*not get water*).

Such a comparison between the dialogues performed by this group of students for OQ1 and OQ3 can be interpreted in several ways. It would be tempting to conclude that the emotional distance to the problem, being a discursive construct on the cognitive plane, merely reflects the degree of students' emotional investment in the discussion. However, these data do not give us access to the students' actual internal emotional state. They only provide us with an image of this state, as expressed in their speech and non-verbal behavior. My hypothesis is rather that the emotional *schematization of* the problem contributes to the argumentative orientation of the debate, producing restrictions on the potential following turns in the interactional sequence, notably in terms of investment. It is difficult to yawn at someone who is raising a life-or-death issue, no matter how sincerely concerned one actually feels about the subject at hand.

• *Case study:* dispute *and high emotional framing*
Another example of the relationship between the social and cognitive functions of emotions in a group argumentation is provided by an emblematic case of *dispute* from the French corpus. Here, it seems, conversely, that it is a too high *thymical* tonality that makes it difficult for the students to cognitively explore the too 'hot' discourse objects. This would lead them to shift the cognitive tension to the social plane, and to deal only with this aspect of the socio-cognitive conflict. Klara, Samira, Isabelle and Asa start disputing in OQ2, and keep on developing *disputational talk* as they apprehend OQ3. They finally manage to soften the conflict and organize a de-escalation at the end of the activity, which allows them to avoid typical *disputate* during the final group dialogue on the main question. However, this de-escalation

mainly involves a strategy of disinvestment: they resolve their socio-cognitive conflict solely on the social plane, without looking for ways of conciliation on the cognitive plane. This observation corroborates the commonly shared idea in the field of collaborative learning according to which effective group collaboration requires an optimal alternation between phases of tension and relaxation. Here, the *thymical* tension appears to be such that it does not allow for sufficient relaxation for the group to engage in cognitive exploration of the problem. Instead, the problem is solved by avoiding the vexing issue, and disinvesting the task, a strategy that is even momentarily made explicit, thematized at the meta-discursive level. Let's look at this group of students' discussion on OQ2 about the efforts they would be ready to make to save water(cf. fig. 6), transcribed below.

1. MOD &changer moins souvent d'téléphone portable et d'ordinateur\ xx xx par rapport à [xxxx (&changing cell phones and computers less often\ xx xx regarding to [xxxx)

2. ASA   [°c'est la A\ nan\° ([°it's A\ no\°)

3. KLA   °nan jamais moi\° (°no i would never myself\°)

4. ASA   °nan c'est la F\ mettez la F\° (°no it's F\ put F on\°).

5. SAM   °<((to Klara)) bah arrête de faire euh> <((looking at the slide, laughing)) de faire aucun effort/>° (°<((to Klara)) bah stop doing uh> <((looking at the slide, laughing)) making no effort/>°)

6. PRO   s'il vous plaît\ [ch:t eh oh\ (please [ch:t eh oh)

7. KLA   [°<((turning to the center of the table)) pff moi j'dis F\>° ([°<((turning to the center of the table)) pff i say F\>°)

8. ANI   faire tout cela et même d'autres efforts\ ne faire aucun de ces efforts\& (do all of this and even other efforts\ not do any of this efforts\&)

9. ISA   [°<((à KLA)) pas F quand même\>° ([°<((to Klara)) not F\>°

10. ANI [&donc euh vous allez redébattre en [p'tit groupe et à trois vous montrerez votre euh: réponse\ allez-y\ ([&so uh you're going to have another discussion in [small groups and on the count of three you'll show your uh: answer\ go ahead\)

11. KLA   [°mais moi j'fais pas l'C hein\° ([°but i do not do the C eh\°)

12. SAM <((seeing a card fall)) ah mais> (<((seeing a card fall)) ah but>)

13. PRO [((picking it up))

14. KLA [moi c'est F hein\ (i am F eh\)

15. SAM <((to the teacher)) merci> (<((to the teacher)) thank you>)

16. ISA   moi j'mets la E hein\ (i put E\ eh\)

17. KLA   oh non moi j'fais F\ (ah no i'm doing F\)

EVT   ((someone comes to pick up the phone that is on the table))
(0.6)
18. SAM ((puts the F on the stand))
(1.6)
19. SAM elle nous a bien enregistrées\ (she recorded us well\)
20. KLA <((looking at the person)) maint'nant r'garde elle est en train d'les enregistrer et tout\ (<((looking at the person)) now see she's recording them and everything\)
21. KLA <((to te center of the table)) moi j'mets la F\ moi j'fais aucun des efforts hein\ c'est pas une personne qui va tout changer hein\> (<((to te center of the table)) i put F\ i don't make any effort eh\ it's not one person who will change everything eh\)
(1.3)
22. ISA bah si moi c'est la E\ y'a plusieurs personnes qui changent/ [ça va changer\ (well if to me it's E\ there are several people who change/ [it's gonna change\)
23. KLA
[oui mais moi j'en f'rais pas partie\ ([yes but i wouldn't take part to it\)
24. SAM bon j'ai mis F hein démerdez–vous\ (so i put F eh just deal with it\)
25. ISA <((after looking towards T5)) elle est grave oh> <((after looking towards T5)) she is crazy oh>
26. KLA ouais\ (yeah\)
(1.2)
27. KLA <((looking at Samira)) toi aussi tu mets F/> (<((looking at Samira)) you also put F/>)
28. ISA <((making a face)) ça s'fait pas: [mais c'est pas F:\ c'est E\> (<((making a face)) you cannot do this: [but it's not F:\ it's E\>)
29. KLA   [ah nan nan mais euh (ah no no but uh)
30. SAM ((removes the F card from the stand))
31. KLA nan mais ça <((hand gesture pointing at Samira and herself)) c'est notre avis à nous deux\> <((hand gesture pointing at the two other students)) vous euh mettez c'que vous voulez\> (no but it <((hand gesture pointing at Samira and herself)) that's our opinion of the two of us\> <((hand gesture pointing at the two other students)) you uh put whatever you want\>
32. SAM ((puts the F card back on the stand))
33. PRO vous vous avez envie d'faire aucun effort alors/ (you you don't wanna make any effort then/)
34. SAM bah [j'pense que nan mais si une personne= (well [i don't think so but if a person=)

139

35. ISA   [mais: elles: sont bêtes hein\ ([but: they: are silly huh\)

36. ASA =nan mais elles sont <((raising hands as beat gestures)) débiles à un point mais pas possible\> (=nan but they are <((raising hands as beat gestures)) dumb to a point but not possible\>)

37. KLA <((hand towards her as saying "stop")) ça va asa hein\> (<((hand towards her as saying "stop")) take it easy asa huh\>)

38. ISA <((hand to ASA)) ouais\> (<((hand to ASA)) yeah\>)

39. SAM nan mais (no but)

40. PRO parle correctement (speak correctly)

41. SAM nous on a notre avis [après vous avez l'sien\ (we have our opinion [after you have hers\)

42. ASA   [j'leur parle pas correctement là ([i'm not talking to them correctly)

43. ISA bah oui mais l'truc c'est [qu'on doit voter pour l'groupe\ (well yes but the thing is [that we have to vote for the group\)

44. KLA   [nan mais j'sais pas si y'a que genre deux personnes qui s'y mettent 'fin c'est pas deux personnes qui vont tout changer quoi\ ([no but i don't know if there's only two persons who start doing it well it's not two persons who will change everything\)

45. ISA bah si c'est plu – [si c'est (well if there's mo– [if it's)

46. KLA   [attends ([wait)

47. ASA   [mais qui vous avait dit qu'y a que deux personnes [qui vont faire (des efforts)/ ([but who told you that there'd only be two people [who will make (some effort)/)

48. ISA
[si c'est oh mais plusieurs personnes ça va changer\ ([if it's oh but several persones it's gonna change\)

49. SAM mais y'a des gens ils sont <((shows the phalanges)) feignants\> regarde on ça fait:& (but there are people who are <((shows the phalanges)) lazy> look we it's been:&)

50. KLA ben ouais mais moi j'le f'rai pas\ (well yeah but i won't do it\)

51. SAM &d'puis [j'sais pas combien d'temps on dit la pollution prenez& (&since [i don't know how long we say the pollution take&)

52. ASA   [tu xx pas les gens/ ([you don't xx people/)

53. SAM &<((with fist beats going down)) les transports [communs et ils continuent à faire des véhicules et ils en achètent\> (&<((with fist beats going down)) public transports [and they keep on making vehicles and they buy them\>)

54. KLA   [ouais ouais c'est vrai\ si c'est encore plus qu'avant\ ([yeah yeah it's

140

true yeah it's even more than before\)

55. SAM <((raising then dropping her hands)) et ça change rien\> (<((raising then dropping her hands)) and it doesn't change anything\>)

56. KLA ouais\ (yeah\)

57. ASA on s'appelle [pas tous samira et et klara hein\ (we are not [all called [samira and and klara huh\)

58. PRO [vous avez pas l'air d'accord alors/ samira j'entends c'que tu dis\ ([so you don't seem to agree/ samira i hear what you're saying\)

59. ASA ((claps hands))

60. PRO tu dis qu'finalement on on nous dit d'le faire mais on l'fait pas [parce que les gens& (you say that finally we we are told to do it but we don't do it [because people&)

61. SAM
[bah oui\ ([boh yes\)

62. PRO [&ils font pas d'efforts\ et <((montre l'écran)) là> <((doigt devant la bouche)) c'est pas> <((touche l'épaule de SAM)) c'qu'on t'demande> c'est pas& ([&they don't make any effort\ and <((showing the screen)) there> <((finger in front of the mouth)) it's not> <((touching Samira's shoulder)) what we're asking you> is not&)

63. ISA [si y'en a qui commencent à l'faire et ben ils le f'ront\ ([if some people start doing it then they'll do it\)

64. PRO &<((hand to the forefront)) c'qu'il faut dire> [<((hand to herself)) c'est c'qui faut faire\> (&<((hand to the forefront)) what you'd say> [<((hand to herself)) that's what you'd do\>)

65. ASA [<((en riant)) (allah aidez-nous\)> ([<((laughing)) (allah help us\)>)

66. ISA ((laughs))

67. PRO est-ce que tu penses qu'il faut l'faire/ [et pas est-ce que tu penses qu'il faut&
(do you think we should do it/ [and not do you think we should&].

68. ISA [<((à ASA, montrant les stylos)) regarde> ([<((to ASA, showing pens)) look>)

69. PRO &l'dire au d — [aux gens d'le faire\ (&tell it to the [people to do it\)

70. KLA [<((turning back, facing the group after looking at the slide)) bah c'est quel effort serais-tu prête/> ([<((turning back, facing the group after looking at the slide)) well that's how much effort would you be ready to/>)

71. PRO voilà\ (here it is\)

72. PRO c'est quel effort <((hand on chest)) serais-tu prête toi\> (it is which effort <((hand on chest)) would YOU be ready to make\>)

73. KLA bah moi c'est la [F hein:\ (boh for me it's [F huh:\)
74. PRO   [toi tu veux faire aucun effort/ ([you don't wanna make any effort/)
75. SAM si:\ [prendre moins d'douches et d'bains\ (yes:\ [taking fewer showers and baths\
76. PRO ((leaves))
77. ASA   [ah: tu m'affiches pas avec ça hein\ ([ah: you don't put that shame on me huh\)
78. ISA   <((grimace)) ah: t'es sérieuse là/> (<((making a face of disgust)) ah: are you're serious/>)
79. SAM et ben: [nan j's'rais pas capable <((pointing gesture at the teacher)) mais he> (and boh: [no i would not be able to <((pointing gesture at the teacher)) but eh>)
80. ASA   [ah: ([ah:)
81. KLA moi <((open hand horizontally placed in front of her)) ce s'rait limite la B mais encore hein\> (to me <((open hand horizontally placed in front of her)) it would be B in a pinch but still hu\>
(1.7)
82. KLA moi qui casse tout l'temps les portables j'pourrais pas les: ne pas les changer hein\ (i who am always breaking cell phones i couldn't not change them right\)
83. SAM attends\ si si si si prendre moins d'douches et [d'vêtements parce que (wait\ yeah yeah yeah yes taking fewer showers and [clothes because)
84. ISA   [nan   moi j'dis E hein\ [E\ mais E:\   ([no I do say E hey\ [E\ but E:\)
85. SAM
[euh: vêtements parce que nan nan parce que r'gardes   ([uh: clothes because no no because look)
86. ISA [faire tous ces efforts–là plus d'autres\ ([making all these efforts and even others more\)
87. ASA [la honte\ on va prendre moins d'douches et de de de de trucs pour économiser d'l'eau hein\ ([what a shame\ we gonna take fewer showers and of of of of things to save water hu\)
88. KLA bah m — moi c'est F\ (bah m - to me it's F\)
89. ISA E (E)
90. KLA <((moving   her   hand   towards   the   others)) débrouillez-vous\> (<((moving her hand towards the others)) you juste handle it\>)
(1.2)
91. SAM pourquoi E/ (why E/)
92. ISA parce que (.) <((showing the slide)) si tu fais tous ces efforts–là> <((raising her hand to a specific heigth)) déjà ça économise> et si <((moving her hand in circle and then putting it back to the previous heigth)) t'en fais d'autres qui [utilisent

142

moins d'eau> ben ça économise encore plus de&   (because (.)
<((showing the slide)) if you make all these efforts> <((raising
her hand to a specific heigth)) first it saves> and if <((moving
her hand in circle and then putting it back to the previous
heigth)) you make others that [use less water> well it saves
even more&)
93. SAM [<((looking for the E card)) E\>   ([<(((looking for the
E card)) E\>)
94. ISA   [&encore de l'eau\   ([&even more water\)
95. KLA [t'as mis quoi toi nadège/   ([what did you put you
nadège/)
96. SAM <((while putting the letter E)) moi j'pense c'est F
parce que [ça sert à rien\>   (<((while putting the letter E)) I
think it's F because [it's useless\)
97. KLA
[tu mets quoi toi/   ([what are you putting you/)
98. ISA   c'est E\ [<((to asa)) tu penses quoi toi/>   (it's E\
[<((to asa)) what do YOU think/>)
99. KLA   [mais <((turning back to T2)) regardes>
<((turning to the outside again)) vous aussi vous mettez F/ (.)
moi j'mets F\>&   ([but <((turning back to T2)) look> <((turning
to the outside again)) you also you're putting F/ (.) me I do put
F\>&)
100. ASA[<((to ISA)) t'arrives pas à comprendre que euh: y'a
(.) <((circle with her hands)) le monde il est entouré de d'vous>
donc euh [si mademoiselle klara et mademoiselle samira
décident&   ([<((to ISA)) you can't understand that uh: there's (.)
<((circle with her hands)) the world is surrounded by people
like you> so uh [if miss klara and miss samira decide&
101. KLA[&<((to another table)) fin moi je j'f'rai aucun effort\>
([&<((to another table)) so i'll make no effort\>
102. ISA [ouais: xx x xxxx\   ([yeah: xx x xxxx)
103. ASA[&de ne pas faire d'efforts   ([&to make no effort)
104. SAM   [nan parce que c'est xx\ r'gardes ça fait j'sais
pas combien d'temps on dit arrêtez d'prendre les [véhicules là
([no because it's xx\ look it's been i don't know how long we've
been saying stop taking [vehicles there)
105. ISA [mais si si bah si on les arrêtait si les si ils on les
écoutait [si ils arrêtaient c'est qui&   ([but if if boh would we
stop if the if they would we listen to them [would they stop who
would&)
106. SAM
[mais ils arrêtent pas\   ([but they don't stop\)
107. KLA[<((to another table)) moi j'mets hein/   ([<((to
another table)) I'm putting on eh/)
108. ISA [&<((opening hands, palms up)) qui aurait [xx c'est
qui s'rait moins pollué/> <((pointing at herself)) c'est nous>

[au bout du   ([&<((opening hands, palms up)) who would have
[xx that's who would be less polluted/> <((pointing at herself))
it's us> [at the end of the)
 109. ASA
[<((holding her ear)) aïe>   ([<((holding her ear)) ouch>)
 110. ISA ((looks at ASA with a smile))
 111. ISA [xxx
 112. SAM   [nan c'est xx justement personne n'arrête\
([no it's xx precisely no one stops\)
 113. KLA[<((to another table)) moi c'est mon état d'esprit\
c'est pas nous qui changerions l'mon[de t'sais euh:>   ([<(((to
another table)) me it's my state of mind\ it's not us who would
change the worl[d you know uh:>)
 114. ISA [mais oui bah [faut commencer\   ([but yes well [you
should begin\)
 115. SAM   [mais c'est ça\   ([but this is it\)
 116. KLA[<((to another table)) vingt ans qu'à la télé ils disent
x la pollution&   ([<(((to another table)) twenty years that on TV
they say x pollution&)
 117. ISA [faut commencer [<((pointing chin at SAM)) (faut
qu't'essaies xxx)>   ([you must start [<((pointing chin at SAM))
(you must try xxx)>)
 118. ASA   [<(((touching   isabelle's   chin))
(arrête) tout l'temps\>   ([<((touching isabelle's chin)) (stop) all
the time\>)
 119. KLA   [&c'est d'pire en pire\ <((shrugging)) personne
fait des efforts\>   ([it's worse and worse\ <((shrugging))
nobody makes any effort\>)
 120. SAM   [mais c'est ça:\ mais ça fait cinq ou six ans
qu'on dit <((dropping her hand on the table)) arrêtez d'prendre
[les voitures:\>   ([but this is it:\ but we've been saying for five
or six years <((dropping her hand on the table)) stop taking
[cars:\>)
 121. ISA [ça ça a pas changé/   ([that hasn't changed/)
 122. SAM   et ils en fabriquent [de plus en plus des
voitures\   (and they are making [more and more cars\)
 123. ISA   [y'en a qui commencent à arrêter
si ils f'raient pas [ils f'raient pas les voitures\   ([some people
start to stop if they would not [they would not make cars\
 124. ASA
[mais ils mettent pas les voitures qu'ils éco — qui qui
fabriquent maint'nant c'est des voitures économiques j'crois\
([but they don't put the cars they eco – that that they make now
are economical cars i think\)
 125. ISA oui\   (yes)
 126. SAM   ouais mais: certaines nan\   euh: (yeah but:
some no\ uh:)

127. ASA <((pointing at ISA)) écologiques plutôt\> [économiques euh: ça l'fait pas trop (<((pointing at ISA)) ecological rather than economical\> [economical er: it's not the word

128. ISA [et si et si tu fais toutes les voitures écologiques [si on arrête de faire [(des xxxx) ([and if and if you make all the cars ecological [if we stop making [(xxxx))

129. KLA [<((turning back to the group)) bon mettez (.)> <((touching the stand)) E/> ([<((turning back to the group)) so put (.)> <((touching the stand)) E/>)

130. SAM <((trying to get the thing asa is playing with)) c'est à moi ça> (<((trying to get the thing asa is playing with)) this is mine>)

131. ASA <((slapping her hand)) dégage\> (<((slapping her hand)) buzz off\>)

132. ISA <((tapping her pen in front of SAM as to get her attention)) c'est oh si on arrête [de faire le truc avec euh:> (<((tapping her pen in front of SAM as to get her attention)) it's oh if we stop [doing the thing with uh:>)

133. ANI [tout l'monde/ ([everyone/)

134. SAM ((grabs what is in asa's hand))

135. ANI [donc euh: ([so uh:)

136. KLA [oh [taisez-vous ([oh [shut up)

137. ISA [<((raising her hand)) écoutes-moi ouaich\> ([<((raising her hand))) listen to me hey\>)

138. SAM j't'écoute\ [si on arrête de faire des trucs avec (i'm listening to you\ [if we stop doing things with)

139. KLA [oh taisez-vous taisez-vous y'a la meuf qui [arrêtez d'parler y'a la meuf qui parle ([oh shut up shut up there's the chick who [stop talking there's the chick who's talking)

140. ISA
[avec l'es — l'essence j'suis désolée mais on peut même il il en restera d'l'essence hein t'auras économisé\ ([with the g- the gas i'm sorry but we can even there there will be gas left eh you'll have saved it\)

141. ANI c'est bon vous vous êtes tous mis d'accord/ (it's good you all agreed/)

142. KLA nan\ (no\)

143. SAM [bah r'gardez si on f'sait des ([boh look if we do)

144. ISA [on fait des éco- ([we do eco-)

145. ASA on fait des voitures [écologiques (we make [ecological cars)

146. SAM [((takes the pen from ASA's hand))

147. KLA [<((to the table behind them)) fin MOI j'suis pas d'accord [avec elles\> ([<(((to the table behind them))] well I don't agree [with them\)

148. SAM [(ça s'rait plus cher\) ([(it would be more expensive\)

149. ISA ouais: mais elle a cru qu'c'était <((circle with her hands around her head)) l'monde [(entier)> ([yeah: but she thought she was <((circle with her hands around her head)) the (whole) [world>

150. ISA ((turning back))

151. ASA c'est d'plus en plus cher mais c'est des voitures [écologiques\ (it's more and more expensive but it's ecologi[cal cars\)

152. SAM
[ben oui je sais nan mais j'suis [toujours pas d'accord c'est tout\ ([yes i know but no i still [don't agree that's all\

153. KLA [F\ (.) nous on n'est pas d'accord/= (F\ (.) we don't agree/=)

154. ANI =c'est bon/ (=that's good/)

155. ISA <((laughing)) toi tu cherches pas l'embrouille> (<((laughing)) you you're not looking for trouble>)

156. KLA j'm'en fous\ (i don't care\)

157. ANI s'il vous plaît/ (please/)

158. SAM <((laughing)) nous on n'est pas d'accord klara\> [on va <((putting her fist on table)) débattre\> (<((laughing)) we do not agree klara\> [we're going to <((putting her fist on table)) debate\>

159. ANI [s'il vous plaît/ faudrait vraiment qu'on passe qu'on coupe le débat en: en équipe si on veut vraiment [que ([please/ we should really move cut the tea: team debate if we really want [that)

160. ISA
[<((to KLA)) r'garde> ([<(((to KLA)) look>)

161. ANI c'est bon [vous vous êtes tous mis d'accord/ (it's good [you all agreed/)

162. ISA [°elle a mis F elle aussi\° ([°she put F her also\°

163. SAM <((levant la E)) ouais\> (<((raising E)) yeah\>)

164. ANI [aller\ euh montre euh: xx ([go ahead\ uh show uh: xx

165. KLA [y'a tout l'monde qui met F\ ([everyone is putting F\)

166. ISA [<(((to KLA, looking at another table)) r'garde xx> [<(((to KLA, looking at another table)) look xx>

167. SAM [<((pointing with her hand)) ils ont mis F tu vois> r'gardes ils ont mis F\ ils ont eu regarde ils ont mis F\ ([<((pointing with her hand)) they put F you see> look they put

146

F\ they got look they put F\)

168. KLA <((taking the card from SAM's hand and giving it to ISA)) nan c'est pas toi qui l'soulèves toi t'es pas E c'est vous qui [l'soulevez>   (<((taking the card from SAM's hand and giving it to ISA)) no YOU don't hold it up you're not E YOU [hold it up>

169. SAM   [<(((en levant le F)) nous on est F\ [klara nous on est F\   ([<(((pulling letter F up)) we we are F\ [klara we are F\)

170. ISA [<(((trying to move samira's hand down)) nan: mets-en pas deux\>   ([<(((trying to put samira's hand down)) no: don't put two\>)

171. ANI [vous pouvez écouter s'il vous plaît/   ([you can listen please/)

172. ISA [<(((trying to lower the F)) mets-en pas deux\>   ([<(((trying to lower the F)) don't put two\>)

173. SAM   [on n'est pas d'accord\ on n'est pas d'accord\   ([we don't agree\ we don't agree\)

174. ASA bah vous êtes pas d'accord vous allez vous faire [foutre <((en riant)) c'est tout\>   (well you don't agree [fuck you <(((laughing)) that's all\>)

175. ISA [(((puts the E on the stand making it fall))

176. ANI alors euh: y'a trois personnes [d'entre vous qui ont répondu: F\   ([so uh: there are three [of you who answered: F\)

177. ASA   [(((laughs))

178. SAM   <((raising letter F)) nan\> [nous on est deux en fait là on on   (<(((raising letter F)) no\> [we are two in fact here we we

179. KLA   [un deux: trois   ([one two: three)

180. ANI bah posez les deux comme ça si vous voulez vous pourrez argumenter [xx   (well put the two this way if you want you will be able to argue [xx)

181. SAM   ((tries to put the F on the stand that ISA has just stabilized and makes it fall again))

182. ISA [t'es pas chiante toi\ [you're a pain aren't you\

183. SAM   ((laughs))

184. ISA x xx faire tomber\   (x xx making it fall\)

185. SAM   mais nan [mais nous on n'est pas [d'accord\   (but no [but we do not [agree\)

186. ASA   [mais moi j'suis pas d'accord pour le E [non plus hein\   ([but I do not agree with E [either hu\)

187. ANI   [y'a: ([there's:)

188. ISA ((looks at ASA))

189. SAM   ((laughs))

147

190. ISA <((trying to put the letters back on the stand)) nan mais toi> (<((trying to put the letters back on the stand)) no but you>
191. ISA ((laughs))
192. ASA ((laughs))
193. KLA ch:t arrêtez arrêtez\ (ch:t stop stop\)
194. ISA °<(((laughing, to ASA)) ça sent la guerre>° (°<((laughing, to ASA)) it smells like war>°)
195. ASA [(((laughs))
196. SAM [(((laughs))

Here we find the characteristics of *disputational* talk: repetitions rather than justifications (negative indicator 1), as in turns 2–11 and 14–17 for example, a limited topical alignment (intermediate indicator 2), very strong but non-constructive critical thinking (indicator 3), and no search for collective assent in the decision-making (negative indicator 4). Students simply repeat proposals and counter-proposals, without explaining the reasons for choosing or rejecting an option, which is essential for a constructive discussion. When a final decision has to be made, Samira decides individually to put the letter F on the desk, in turn 24, indicating to the others that she does not care about their opinion: "so i put F eh just deal with it\". Similarly, Klara, who agrees with Samira, openly shows her disinterest in what the other two girls think, her decision being already made (turn 31). Much later, when the facilitator tells them that it is time to raise the cards representing the chosen options, Samira initiates a de-escalation strategy by de-investing the task, as she raises the letter chosen by Isabelle and Asa, corresponding to the option rivaling her own, just to get rid of the problem (turn 163). Finally, this decisive moment does not involve the search for a collective agreement, but gives rise to conflicting gestures: snatching a card from the hands, forcing the other to put down her hand (turns 168–170).

The fifth and final indicator is also negative. In the whole-class discussion that follows this dialogue, the students in this group speak only twice to mention what was said in the small group, always using only their own initial ideas. The other students in the group simultaneously criticize what is being said, in a low voice, to show their disagreement, and to destabilize the speaker. When Samira addresses the class, Asa whispers to her:

ASA °tu mets pas ma bouche dedans° (°don't put my mouth in it°)

This typical case of disputational talk correlates, for both OQ2 and OQ3, with a very intense thymical tonality (Polo et al., 2016). Table 19 summarizes the data from this dialogue, corresponding to two parameters framing emotional intensity: the concerned persons and the responsible agents. No statements in this group dialogue specifically addresses spatiotemporal distance.

Table 19. Elements constructing the emotional distance to OQ2 in the studied group dialogue.

Emotional tonality: distance to concerned and responsible people in group dialogue on OQ2	
Concerned people	ISA c'est qui s'rait moins pollué/ <((pointing at herself)) c'est **nous**> (that's who would be less polluted/> <((pointing at herself)) it's **us**>)
	KLA °nan jamais **moi**\° (°no **i** would never **myself**\°) SAM °<((to Klara)) bah **arrête** de faire euh> <((looking at the slide, laughing)) de faire aucun effort/>° (°<((to Klara)) bah **stop** doing uh> <((looking at the slide, laughing)) making no effort/>°) KLA °**moi** j'fais pas l'C hein\° (°but **i** do not do the C eh\°) KLA **moi** j'fais aucun des efforts hein\ c'est pas une personne qui va tout changer (**i** don't make any effort eh\ it's not one person who will change everything ISA y'a **plusieurs personnes** qui changent/ ça va changer\ (there are **several people** who change/ it's gonna change\) KLA **moi** j'en f'rais pas partie\ (**i** wouldn't take part to it\)
	SAM si **une personne** (if **a person**) KLA c'est pas **deux personnes** qui vont tout changer (it's not **two persons** who will change everything) ISA **plusieurs personnes** ça va changer (**several persones** it's gonna change) SAM des **gens** ils sont <((shows the phalanges)) feignants\> (**people** who are <((shows the phalanges)) lazy>) KLA **moi** j'le f'rai pas\ (**i** won't do it\)
Agents responsible for the evolution of the situation	ASA les **gens** (**people**) SAM **ils** continuent à faire des véhicules et **ils** en achètent (**they** keep on making vehicles and **they** buy them) ASA on s'appelle pas tous **samira** et et **klara** (we are not all called **samira** and and **klara**) ISA qui **commencent** à l'faire (**start** doing it) SAM j's'**rais** pas capable (**i** would not be able to) KLA **j'pourrais** pas les: ne pas les changer (**i** couldn't not change them) ASA **on** va prendre moins d'douches (**we** gonna take fewer showers) ISA si **tu fais** tous ces efforts-là (if **you** make all these efforts) ASA le monde il est entouré de d'**vous** (...) **mademoiselle klara** et **mademoiselle samira** (the world is surrounded by people like **you** (...) **miss klara** and **miss samira**) KLA **j'**f'rai aucun effort (**i'**ll make no effort) ISA si **on** les arrêtait (...) **on** les écoutait (would **we** stop (...) would **we** listen to them) SAM **ils** arrêtent pas (**they** don't stop) SAM **personne** n'arrête (**no one** stops) KLA c'est **pas nous** qui changerions l'monde (it's **not us** who would change the world) ISA (faut qu'**t'essaies**) (**you** must try)

KLA **personne** fait des efforts   (**nobody** makes any effort)

ISA **y'en a qui commencent** à arrêter   (**some people** start to stop)

ISA si **on** arrête   (if **we** stop)

SAM si **on** f'sait   (if **we** do)

However, the students largely continue their dialogue in parallel with the whole–class debate that follows this group discussion. Table 20 lists all the elements of this side dialogue that also contribute to the construction of the *thymical* tonality, particularly in terms of emotional intensity. We can see that major characteristics emerge clearly, both in terms of the concerned and responsible people and of the spatial and temporal distance from the problem of accessing to drinking water.

Table 20. Emotional distance constructed during the side dialogue on OQ2.

Emotional tonality: construction of the distance to OQ2 during the side dialogue		
Utterance		**Features**
KLA	ça **va pas** changer qu' y ait **deux ou trois personnes** (it's **not** gonna change would there be **two or three people**)	near future, others' responsibility
SAM	y'aurait **moins que la moitié** (there'd be **less than half**)	others' responsibility
ISA	trois millions d'personnes ça **va changer** (three-million people **it's gonna** change)	near future, others' responsibility
KLA	y'en **aura mille** alors (there'**ll be a thousand** then)	future, others' responsibility
KLA	si **on** est **trois millions ça changera** (if **we** are **three million people it will change**)	future, own responsibility
ASA	**nous** chronométrer dans la douche (to time **us** showering)	first person both as responsible and victim
ISA	si y'a **plusieurs millions d'personnes** (if there is **several million people**)	others' responsibility
ISA	s'**ils** le f'raient (would **they** do it)	others' responsibility
SAM	c'est **la nouvelle génération on** est pourri gâté (it's **the new generation we** are spoiled)	present time, own responsibility
SAM	à **c't'époque** (in the **current time**)	present time
KLA	va trouver la foi en **les gens** (go find faith in **the people**)	others' responsibility
ISA	y'en a **plein** qui le font (there are **many people** who do it)	present time, others' responsibility
SAM	**on** pense pas\ **on** pense à la vie\ (**we** don't think\ **we** don't think about life\)	present time, own responsibility
KLA	**moi mes** besoins j'les f'rai (**me my** needs I'll do them)	future, own responsibility
SAM	**ma** douche **j'**y passe trente minutes **j'**la ferai (**my** shower I spend thirty minutes in it I'll do it)	present time & future, own responsibility
ASA	s'chronométrer dans la douche\ (timing **oneself** showering)	concerned people : others

ASA	j'me réveille dix minutes tulutulu (i wake up ten minutes tulutuluà	concerned people : me
KLA	tu vas pas faire ça hein\ ça sert à quoi (you won't do this hu\ what for)	second person both responsible and victim
SAM	y'a du savon tu sors quand même (there's soap you get out anyway)	second person both responsible and victim
SAM	mais à autrans euh nan j'sais plus quoi en russie ils étaient chronométrés deux minutes par douche\ (but in autrans er no i don't know any more in russia they were timed two minutes per shower)	social and spatial distance
SAM	ils avaient moins d'cinq minutes (they had less than five minutes)	concerned people : others
SAM	en russie vous aviez moins d'cinq minutes (in Russia you had less than five minutes)	spatio-temporal distance, concerned people: second person
KLA	t'es habituée à faire un truc tu vas pas changer (you're used to doing something you're not gonna change)	own and everyone's responsibility
KLA	bah c'est la personne que t'es (well it's the person who you are)	second person both responsible and victim
KLA	tu vas pas arriver avec en pat d'eph au bahut (you're not gonna come to school with old pants)	second person both responsible and victim
ISA	si les gens (if people)	others' responsibility
KLA	les pubs qu'ils font (the ads they do)	others' responsibility
ASA	s'chronométrer dans la douche\ (timing oneself showering)	concerned people : others
ISA	qu'est-ce que j'f'rais sans mes bains moi (what would i do without my baths)	concerned people : me
KLA	notre hygiène de vie (our life hygiene)	concerned people : us
ISA	on va prendre les conséquences de nos actes (we gonna take the consequences of our actions)	near future, us as both responsible and victims
KLA	moi j'vais pas changer pour les autres (I'm not gonna change for others)	concerned people: others, responsible: me
ASA	pour toi ta vie ta santé la santé de tes enfants (for you your life your health your children's health)	present time & future, victim & responsible: you
ISA	enfants p'tits-enfants (children grand-children)	future, concerned people: you (your beloved others)
KLA	j'vais pas changer mes habitudes à m'faire chier à (i'm not gonna change my habits gonna bother to)	first person both as responsible and victim
SAM	on va pas mettre un chronomètre (we're not gonna put a chronometer)	first person both as responsible and victim
KLA	faire chier mes parents (bothering my parents)	victim and responsible: me (my beloved others)
KLA	au bout d'un moment j'le f'rais si y'en a vraiment b'soin (at some point I'll do it if it's really needed)	future, own responsibility
SAM	t'es plein d'savon tit tit il faut qu'je sorte là ça a sonné\ (you're full of soap tit tit i must get out it rang)	first person both as responsible and victim
KLA	moi j'change pas mon mode de vie pour l'eau (i do not change my lifestyle for water)	first person as responsible, no victims

ASA	les efforts c'est qu'**les gens ils** arrêtent de mentir\ (the efforts is that **the people** stop lying)	others' responsibility
ISA	**t'**arrêtes de laver **ta** voiture (**you** stop washing **your** car)	own and everybody (second person)'s responsibility
SAM	arrêter d'laver **sa** voiture (stop washing **one's** car)	everyone's responsibility
KLA	**moi j'**le fais pas (I don't do it)	own responsibility
SAM	**tu vas** laver avec (**you**'re gonna wash with)	near future, own and everybody (second person)'s responsibility
KLA	tant qu't'es pas allée **dans l'futur** (as long as you did not go to **the future**)	future

At first glance, there is a contradiction between the way the students present the concerned people and their spatio-temporal framing of the problem. On the spatio-temporal level, the problem is depicted as distant from them: it affects Autrans, Russia... Similarly, the students present the situation as likely to become critical in the future, and not as immediately worrying. However, we can observe *phasic* variations on this temporal distance: sometimes it is a question of a very near future, implying that the students may themselves suffer the "consequences of their actions" if they waste water; sometimes of a more distant future likely to affect future generations, their "grand-children". The present is referred to only as the time when the actions causing the water scarcity problem are committed. Conversely, the students present, in their discourse, the people concerned by the problem of access to water as very close to them. They often use the first person, singular or plural (*I, we*), and the second person, mainly singular (*you*), with a plural form. Even when the third person is used, it is associated with emotional proximity. Two occurrences of *they* (when Samira says "they were timed; they had less than two minutes") actually correspond to the *you* used later, to refer to some students in the group who went on a school trip to Russia. Other occurrences of the third person refer to other close relations: the students' future children or grand-children. Only a few occurrences use the third person (*they, others*) referring to distant people. The problem, however, is not depicted as radically serious, likely to lead to death. Most of the time, the concerned people, mainly the students themselves and their families, are characterized as water consumers facing the problem of the price of water as they need large quantity of water to maintain their habits and satisfy their personal comfort.

The contrast thus described between the two parameters of spatio-temporal framing and the concerned people shows that distance from the problem is not solely and purely determined by objective material conditions (such as local drinking water resources) but is indeed the result of a discursive construction. Here, the four students, even though they are aware that they do not belong to the population most threatened by water shortage, make an important effort to take the issue seriously, both as a global concern for humanity and as their own problem.

Finally, these students also *schematize* the agents responsible for changing the situation of access to drinking water. There is a lot of talk about this, with a strong identification of the students themselves as agents potentially responsible for improving the situation. This identification is made through the extensive use of the first person, in the singular (*I, me*), or in the plural (*we, us*). It is also elaborated through the extensive use of the second person singular, with the students accusing each other of being responsible for the situation (with even direct interpellations using the first names of Samira and Klara). Interestingly, such alternation of first and second person singular is a discursive mark of conflict in French (Denis, et al. , 2012), which is consistent here with the analysis of the dialogue as *disputational* talk. The other depictions of those in charge of the situation use the third person, with a variety of meanings. Several occurrences of "they" refer to the precursor group that would initiate a change in practice that would reverse the trend and preserve access to water. There is some debate among the students about the minimum number of people needed for this group to make a difference, and whether or not it is worth getting personally involved in this cause. When it comes to the latter, the third person sometimes has a first person value, as long as students include themselves in this precursor group. Isabelle's statement, during the side dialogue about OQ2, is emblematic of this phenomenon: "if there are three million of us it will change". Another group of statements in the third person refers to a general entity that does not explicitly involve the students, but from which they cannot totally disassociate themselves: "the people". This relatively indeterminate global agent is presented as resistant to habit change, self-centered and lazy, and representing the majority of society. Many of these occurrences are embodied in the term "nobody" a semantically radical form that theoretically also includes the students themselves. Another type of third-person usage refers to people who are seen as the most responsible, the ones to be blamed for the situation, the people taking advantage from the fact that access to water has become or is likely to become problematic. Students do not generally include themselves in this group, which is primarily defined as the set of industries benefitting from the situation. A statement in the third person has the specificity of referring to people considered both as potential victims and agents responsible for the situation: the "parents" whose way of life has led to this situation, a way of life which may however be transformed to change the situation, notably by having to clean dry toilets.

Considering all these parameters, it appears clearly that the emotional *thymical* tonality in this group discussing the OQ2 is very intense. Emotional proximity is constructed by a double identification: with the people concerned by the problem, and with those responsible for the situation and/or its evolution. Such a *thymical* framing favours the use of self- and hetero-accusations, dynamics conducive to feelings of guilt and offence. These emotions typically play a social function in engagement into a *dispute*.

Moreover, during these dialogues the students make many meta-discursive comments on their activity, that they depict rather negatively, with argumentative norms associating any debate with the polemical genre.
As they are asked by the facilitator to raise the letter of their choice for the group vote on the OQ2, they make the following comments about what they are doing:

> ISA on doit voter pour l'groupe (we have to vote for the group) (…)
> ASA   tu m'affiches pas avec ça (you don't put that shame on me) (…)
> ISA <((laughing))   toi   tu   cherches   pas   l'embrouille> (<((laughing)) you you're not looking for trouble>) (…)
> SAM   <((laughing)) nous on n'est pas d'accord klara\> on va <((putting her fist on table)) débattre\>   (<((laughing)) we do not agree klara\> we're going to <((putting her fist on table)) debate\> (…)
> KLA   <((taking the card from SAM's hand and giving it to ISA)) nan c'est pas toi qui l'soulèves toi t'es pas E c'est vous qui [l'soulevez>   (<((taking the card from SAM's hand and giving it to ISA)) no YOU don't hold it up you're not E YOU [hold it up> (…)
> ISA °<((laughing, to ASA)) ça sent la guerre>°   (°<((laughing, to ASA)) it smells like war>°)

Here, both the lexicon and the gestures refer to the semantic field of conflict and war, an analogy explicitly used to describe the current interaction. When the facilitator later announces thegroup discussion on OQ3, Samira expresses her anxiety about a new potential conflict:

> SAM   °on va s'taper là\°   (°we're gonna hit each other now\°)

Similar meta-discursive comments appear when discussing OQ3:

> SAM   bon maint'nant on fait pas d'merde hein   (well now we don't do shit right) (...)
> ASA   ((surrounds her head with her hands simulating a big head)) (...)
> SAM   isa tu défends ta cause   (isa you defend your cause) (...)
> SAM   on est en train d's'entretuer   (we are killing each other) (...)
> ISA c'était d'la merde   (it was shit) (...)
> SAM   tout l'monde se dispute   (everyone is quarrelling) (...)
> ASA   mais c'est bon arrêtez avec vot' débat on va pas parler   (but it's okay so stop with your debate we are not going to talk) (...)
> ISA t'arrêtes de t'exciter toi un peu là/   (you stop a bit getting excited now/)

Such a conception of the activity contributes to the opposition between strong individual beliefs evolving into interpersonal conflict. The students seem to view argumentation only as a threat to their relationships with each other. As a result, the activity is almost only depicted through its social consequences, and almost nothing is said, at the metadiscursive level, about the problem of reconciling or co-elaborating alternative viewpoints on the topic that the group is supposed to address, on the cognitive plane. Ultimately, it is a strategy of cognitive disinvestment that is enacted by these students, reinforcing the group through the reactivation of a shared role of "bad student" that seems familiar to them. This stigma is explicitly mentioned as a shared identity as if it served to (re)build group unity. This is particularly visible in the self-deprecating side discourse produced by the students during the class debate on the OQ2:

> ISA <((looking at the teacher)) madame dupont>   (<((looking at the teacher)) madame dupont>) (...)
> KLA   eh elle va s'dire elle aura honte de notre classe\   (eh she will think she will be ashamed of our class\) (...)
> KLA   aussi ils ont pris la pire classe comme ça euh ils ont pris les pires gens d'la classe   (they also took the worst class this way uh they took the worst people in the class)

- *Articulating the social and the cognitive from emotions: theoretical implications*

These two cases constitute arguments for hypothesizing a relationship between the emotional intensity of *thymical* framing and the nature of group talk. I propose to specify this hypothesis at the theoretical level by completing the model of the functions of emotions in group argumentation. Figure 12 is to be understood as an addition made on a zoom of the lower part of figure 11, representing the cognitive and social results from there role played by emotions during a collective argumentation task. It symbolizes the correlations observed in these two cases between the collective configurations structured by the emotions at the social and cognitive planes, and their effects on the argumentation developed during the debate.

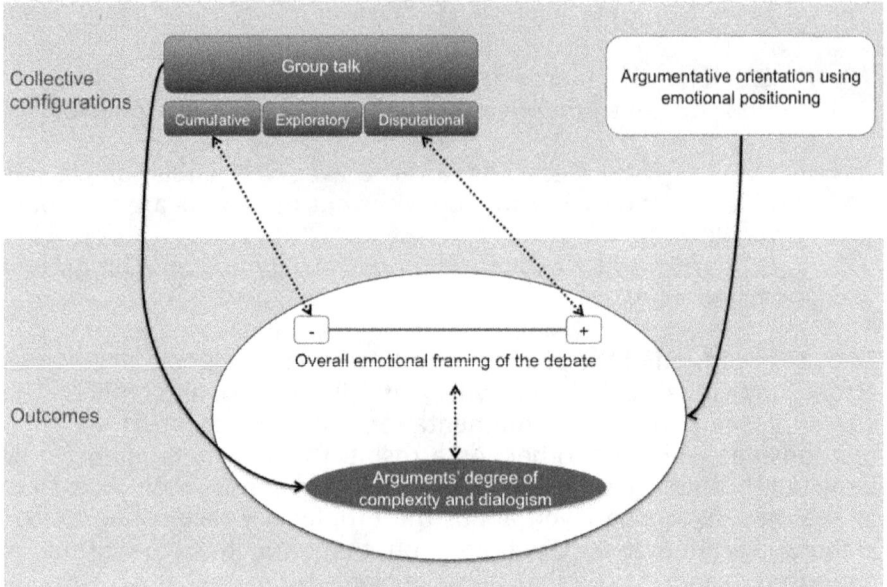

Figure 12. Relationship between *group talk* and *thymical tonality:* complexifying the model of the social and cognitive functions of emotions in group argumentation.

The type of group talk and the *thymical* framing of the debate can be understood as related in the following way: a too low emotional intensity would not allow for sufficient investment to truly explore the issue, being associated with the *cumulative talk;* while a too high emotional intensity, on the other hand, would displace the problem from its cognitive core to its relational effects, favoring *disputational talk.* These interpretations corroborate previous studies in the field of collaborative learning, first on the necessity of a positive minimal tension to stimulate group work (e.g. Sins & Karlgren, 2013), and secondly on the fact that the feeling of offensiveness is likely to inhibit group reasoning (Muntigl & Turnbull, 1998). Note that such links are not unidirectional causal relations, but rather stands as feedback loops. For example, specific emotions playing a social function, such as the feeling of offence, can affect the *thymical* emotional framing by inciting the participants to feel more or less concerned by the problem, or responsible for its evolution. Conversely, the construction of distance to the problem, a key component of the cognitive function of emotions, is likely to produce more or less intense emotions in students depending on whether the problem appears as more or less dramatical to them. Therefore, the relationship between the social and cognitive functions of emotions is represented here as a *continuum*. Indeed, many authentic dialogues cannot be qualified as belonging globally to a type of group talk (cf. 3.1.2). It is likely that they also present different sequences regarding the intensity of the *thymical* framing of the debate, or even hybrid transitional sequences marked by a progressive evolution of this intensity. Therefore, there is no theoretical reason to exclude any potential intermediate situation between these two poles (very intense/very low emotional framing).

- *Articulating the social and the cognitive from emotions: practical directions*

All in all, such theoretical hypothesis points out that the pedagogical tools and strategies aimed at promoting *exploratory* talk cannot ignore the problem of optimizing the emotional framing of the activity. The model of group emotions in argumentation thus has several concrete implications for instructional design.

First, it questions a restrictive view on the matter of emotion regulation, which sometimes leads educators to seek to isolate "pure" cognitive processes from emotions, though as potentially fallacious. In my opinion, it is more fruitful to try to make people aware of the social functions of emotions, and to provide them with tools for collective regulation (e.g. Järvenoja & Järvelä, 2013). The benefits of emotion *awareness* are less obvious at the cognitive level. Indeed, French students in the case of typical *disputational* talk demonstrate very clear awareness of the emotional "heating" of their conflict, but this leads them to purely social relaxation strategies based on cognitive disinvestment. The most interesting way to take into account the socio-emotional dimensions of reasoning is to build an initial pedagogical situation that really takes them into account, together with its cognitive dimension. Understanding the cognitive emotions that support students' engagement in specific types of talk that are more or less favourable to collective reasoning is necessary to design educational tools or activities that enable quality interactions between students. However, this does not necessarily mean supplementing pre-existing pedagogical environments with a few tools for emotional awareness. Indeed, an optimal *thymical* framing can only result from a cognitive process led by the didactic situation itself. Here it comes to the matters of choosing topics of discussion and contextualizing the target knowledge in a way that supports appropriate emotional investment in the group argumentative task. Designers must also take into account these socio-emotional aspects when creating the overall scenario (*scripting*) of the activity (Weinberger et al. 2005).

If facework is so structuring for interactions, it is because of its affective nature. Expressing emotions related to the communicative situation, either verbally or non verbally, in a more or less implicit or explicit way, allows eahc member of a group to adjust his or her self-identity footing with others'. Here, emotions are serving social functions: they mediate interactional alignment towards a specific group talk. Such mechanisms can be compared to the cognitive functions of affects as emotions are used to frame a problem and defend an argumentative conclusion. Individuals then align argumentatively through phasic moves orientating the debate towards their shared opinion. If such a distinction proves useful to understand the role of emotions in group argumentation, it should not be exagerated. Actually, the intensity of thymic tonality and the nature of group talk seem to be related. Two case studies show consistent tendencies: too low intense framing would not motivate cognitive investment enough, resulting in cumulative talk; while too high intense framing may lead to dispute. How can the model of the functions of group emotions in argumentation (cf. fig. 9 et 10) be useful for the practice? It raises the following questions: can we possibly define a thymic framing that would be optimal for collective reasoning and if so, how can we scaffold pedagogical activity to reach it? Is it possible to plan, or to regulate in real-time the emotional framing of a debate? On the social plane, we may help the students engage into a real cognitive exploration of the problem by explicitly teaching them the specific politeness rules of the argumentative task. It is also possible to influence social emotions through role-play, or scripting student grouping (avoiding or favoring existing affinities, combining genders and levels, etc.). But it seems more difficult to imagine didactical strategies to directly influence emotional schematization, on the cognitive plane. Actually, the latter also depends on group and individuals' *ex ante* relation to the topic, notably the what Amossy calls the *interdiscourse* to which they were exposed about the controversy (Amossy, 2006: 94-99). Media and leaders' discourses about the question necesarily structure students' first apprehension of the controversy, both emotionally and cognitively. There, didactical approaches on socially accute questions through media analyses are particularly interesting (e.g. Jimenez-Aleixandre, 2006).

3.3 Attacking the person

This hypothesis of a link between the emotional framing of the problem and the dynamics favourable to collective reasoning is part of the general question of defining what constitutes a socio–affective climate fostering cognitive advancement, and how to bring it into practice. One practice seems to play a fundamental role in this respect: the use of arguments challenging the people partipating in the discussion. Indeed, such a process exploits and relates the three dimensions of argumentation: relational, emotional and cognitive. At the affective level, any argumentative staging, positive or negative, of a person involved in the interaction, intensifies the emotional level of the debate, via the parameter of the social distance to the issue. For example, portraying the person with whom one disagrees as a wealthy landowner taking advantage of the status quo in a discussion about land reform may "warm the exchange up". However, such emotional framing is cognitively part of the problem's schematization, and is not entirely irrelevant to the content of the issue. At the relational level, such a statement is bound to be a source of tension: it is likely to be experienced by the depicted person as a personal offence, a feeling likely to lead him or her to adopt a competitive identity footing. If he or she reacts by offending others in turn, and everyone else aligns themselves with such an identity footing, disputational talk develops.

Argumentation theory has long been pointing at the "danger" of such arguments "attacking the person" (ad personam being considered as a typical example of fallacious argument). Their avoidance even seems to be part of common argumentative norms, a principle often mentioned in ordinary argumentative discourse (Doury, 2004, p.54). Paradoxically, the prevalence of this denunciation of personal accusations also reveals the extent to which such practices are usual, and often identified by the argumentators themselves. It is not surprising that one of the rules presented as fundamental to constructive group argumentation even embodies this principle: putting ideas in competition with each other, not individuals (Asterhan, 2013, p. 254). How come such personal attacks remain so frequent? What functions do they play in argumentative interactions? Based on the study of the YouTalk corpus, I propose a typology of argumentative personal attacks, which accounts for the diversity of practices falling under this general category, identifying their interactional effects. By focusing on the cases of disputational talk of this corpus, I then highlight the forms of personal attacks that tend to be most associated with blocking the discussion. Finally, I make some practical recommendations for creating a socioaffective climate conducive to group learning.

3.3.1 Typology of personal attacks

I identified four forms of argument in the YouTalk corpus that fall under the "personal attack" of others, two that are relatively well-known in standard classifications of arguments and two for which I found no equivalent in pre-existing typologies: the *ad hominem*, the *ad personam*, and what I will call the accusation of sociocentrism, and the accusation of incompetence. The classic *ad hominem* argument can be defined as a rebuttal process where "the speaker rearticulates the system of beliefs and values of the opponent, in order to identify a contradiction" (Plantin, 2018, 'Ad hominem'). The contradiction is unearthed through inherently tension–creating discursive work, which "embarra[sses] the opponent and cau[ses] him or her to reconsider his or her speech, positions or actions". To this end, it is possible to draw "not only from what has been said by the opponent in the past, but also from what has been said by "his or her people", that is to say, by members of the discursive community sharing the same argumentative orientations: people of the same party, religion, scientific trend, etc., that cannot be easily disavowed". Plantin distinguishes between 1) contradictions between words or between words and beliefs; 2) contradictions between words and practices.

The *ad personam* argument, on the other hand, aims to disqualify the speaker in order to disqualify what he or she says (Plantin, 2018, "Personal attack"). Among these procedures, Plantin lists 1) the insult; 2) ironization on some irrelevant aspects of the opponent; 3) the disqualification of the person by mentioning elements of his/her private life that are not relevant to the discussion. The ad personam argument is the form of personal accusation the most often condemned by common standards of politeness or argumentation. It is also condemned by critical theories of argumentation in that this type of argument is not about the substance of the controversy (Ibid., "Matter"). Plantin notes that it is generally a source of emotional intensification of the debate, and of displacement of the latter on the level of identity: "the opponent will be upset, he or she will lose track of the argument and will finally resort in turn to personal attacks and insults".

My empirical work has led me to define another major type of argument attacking the person, the accusation of sociocentrism. It is based on a strong belief in social determinisms in the Bourdieusian sense[13], and consists in discrediting a person's statements as being guided by the interests and specificities of a social group to which he or she belongs. Like the ad hominem argument, it uses the social affiliation of the interlocutor but, rather than contradicting his or her words and the belief system in which they are situated, it instead denounces the "hidden motive", perhaps even unconscious, that drives his or her discourse, for reasons of personal or class own interest constituting a "guilty motive" (Plantin, 2018, "Motives and reasons"). Thus, this argument can be classified as a form of ad personam, in the sense that it discredits the speaker, in order to refute his claims. However, the reasons for this discrediting concern elements that are depicted as relevant to the topic under discussion, and, very often, can hardly be judged as not being about the substance of the issue. This

---

[13] Adjective referring to the work of the French sociologist Pierre Bourdieu, who studied the weight of social determinisms in many dimensions of social life (school, cultural practices, etc.).

argument can be understood as an argument based on the negative reputation ethos[14] attached to the social group mentioned, in a strategy of "presenting the opponent" as untrustworthy[15], due to a lack of concern for the common good (eunoia), and a focus on one's own group interests and values. This tool of rebuttal is condemned by the fifth rule for "honorable controversy" defined by Levi Hedge in 1838: "No one has a right to accuse his adversary of indirect motives." (Plantin, 2018, "Rules"). The accusation of sociocentrism relies on classic topics on the person based on typical socio-rhetorical identity parameters such as family, nation, gender, age, condition, character dispositions, life styles.

Another strategy based on depicting the opponent is sometimes employed by students, presenting him or her as untrustworthy because of a lacking of intelligence. I have called this process the accusation of incompetence. It is similar to the ad ignorantiam refutation, condemned by critical theories of argumentation as an argument that does not address the substance of things, but relies on an attack on the opponent (Plantin, 2018, "Matter"). In a slightly different sense, Locke defines it as admitting to be true what has not been shown to be false, i.e., fallacious reasoning consisting in reversing the burden of proof: if the other does not have sufficient knowledge to show the invalidity of my argument, then it is true (Plantin, 2018, "Ignorance"). I use the term accusation of incompetence here as refuting the opponent's claims by presenting them as meaningless, and only resulting from the opponent's incompetence.

Finally, I have also identified a specific personal attack form, which breaks the argumentative routines, because it does not concern the opponent but oneself: self-deprecation. The rules of the argumentative interaction script would have everyone exploit his or her ethos in such a way as to convince, through positive self-presentation, and/or attempting to discredit the opponents. This dynamic would even characterize a politeness system specific to argumentative interactions (Plantin, 2018, "Politeness"). Therefore it is relatively surprising to find self-deprecating practices in this corpus, either only targetting the individual or about his or her social group, potentially including other members of the interaction. However, this phenomenon can be understood as a marker of identity vulnerability, likely to shed light on possible difficulties in reasoning collectively.
Table 21 presents this typology of forms of personal attacks, illustrating each category with examples taken from the YouTalk

---

[14] What Amossy calls the "prior ethos" (2006, p. 70).
[15] This corresponds to the flip side of self-presentation as conforming to persuasive authority, a typical mechanism one of the argumentative uses of *ethos* (Plantin, 2018, "Ethos").

corpus, reproducing, when possible, one example for each country. The only exception deals with the accusation of incompetence, a category for which I found no occurrences in the American data. This may be due to the fact that there are generally fewer personal attacks in the American corpus than in the other two national subcorpora, with French students being the most prolific in this regards. This is consistent with the fact that there were no typical cases of disputational talk among the American students studied, and that it is the French students who tended to develop the most disputational talk in these data.

Table 21. Ethotic strategies: five forms of argumentative personal attacks.

| Personal attacks | Personal attacks on others | | | | Self-deprecation |
	Ad hominem	Ad personam	Accusation of sociocentrism	Accusation of incompetence	
Definition	Contradiction between statements or statements and beliefs or acts	Irrelevant disqualification of the opponent	Limitation of the opponent's reasoning due to his social position	Limitation of the opponent's reasoning due to ignorance or lack of intelligence	Attacking oneself or one of oneself's social group
French example	elle prend deux trucs qui s'contredit (she takes two things that contradict each other)	t'es pas chiante toi\ (you're a pain aren't you\)	on s'appelle pas tous samira et et klara (we are not all called samira and and klara)	elles sont <((raising hands as beat gestures)) débiles à un point mais pas possible\> (they are <((raising hands as beat gestures)) dumb to a point but not possible\>)	<((talking about herself)) ils ont pris la plus conne de nous quatre> (<((talking about herself)) they took the dumbest one of us four>
Mexican example	lo que comentamos ahorita en la actualidad van a hacer esto/ ustedes lo hacen/ (what we are saying now you are actually going to do it/ you are doing it/)	aï el burro (aï what an idiot)	tú no vas a tener (...) y (...) vas a verlo desde otro punto de vista (you're not gonna have (...) and (...) you're gonna see it from another viewpoint)	están mal\ (you are wrong)	como somos los mexicanos de responsables vamos a: vamos a ahorrar agua/ (as responsible as we are the mexicans we're gonna: we're gonna save water/)

	unless <((pointing at him)) you			
American example	volunteer yourself to work to water free treatment	your hair looks thirsty for some shampoo	alright\ it then you're just being selfish\	if i had the opportunity but i'm not gonna get one

3.3.2 *Dispute*, personal attacks and rhetorical styles

In my dissertation, I conducted an exhaustive inventory of these forms of personal attacks in the YouTalk corpus. The empirical results seem to confirm the hypothesized correlation between *disputational* talk and personal attacks, especially when personal attacks are carried out at the level of small discussion groups (Polo, 2014, pp. 193–199).

The typical case of *dispute* in the French corpus presented above (cf. 3.2.3) illustrates this tendency, since the dialogue between Klara, Asa, Isabelle and Samira is full of personal attacks. Table 2 lists t2hem following the typology of the five forms of personal attacks presented above. It should be noted that, although the *self-deprecations* made concern all the present students, and even the whole class, they are made by two of the four members of the group (Samira and Klara), whereas the personal attacks on others are made by the other two students of the group (Asa and Isabelle). It seems that the students who are the most personally attacked by their classmates integrate this negative perception of themselves, even going so far as to present themselves in their own speech as incompetent to argue.

Table 22. Personal attacks in a typical dispute case from the French corpus.

Form of personal attack	Utterance
Ad hominem: contradiction between words and actions	ASA t'arrives pas à comprendre que euh: y'a (.) <((circle with her hands)) le monde il est entouré de d'vous donc euh si mademoiselle klara et mademoiselle samira décident (…) de ne pas faire d'efforts (you can't understand that uh: there's (.) <((circle with her hands)) the world is surrounded by people like you> so uh if miss klara and miss samira decide (…) to make no effort)
Ad personam	ISA elle est grave oh (she is crazy oh) ISA toi tu cherches pas l'embrouille (you you're not looking for trouble) ISA t'es pas chiante toi (you're a pain aren't you) ISA <((trying to put the letters back on the stand)) nan mais toi> (<((trying to put the letters back on the stand)) no but you>)
Accusation of sociocentrism	ASA on s'appelle pas tous samira et et klara (we are not all called samira and and klara) ASA ouais: mais elle a cru qu'c'était l'monde entier (yeah: but she thought she was the whole world)

Accusation of incompetence	ISA elles: sont bêtes (they: are silly) ASA elles sont <((raising hands as beat gestures)) débiles à un point mais pas possible\> (they are <((raising hands as beat gestures)) dumb to a point but not possible\>)
Self-deprecation	KLA <((talking about herself)) ils ont pris la plus conne de nous quatre> (<(((talking about herself)) they took the dumbest one of us four>) KLA moi j'suis pas écolo en fait\ (I am not ecological in fact) SAM ça changera pas d'la vie d'tous les jours\ que je dise de la merde (it will not change from dayly life\ that i say bullshit) KLA j'sais qu'j'suis têtue (i know that i'm stubborn) SAM c'est la nouvelle génération on est pourri gâté (....) on pense pas à la vie (it's the new generation we are too spoiled (...) we don't think about life) KLA elle aura honte de notre classe (...) ils ont pris la pire classe comme ça euh ils ont pris les pires gens d'la classe (she will be ashamed of our class (...) they also took the worst class this way uh they took the worst people in the class)

How can we interpret this frequency of *disputational* talk and personal attacks in the French corpus? One possible interpretation lies in the cultural dimension of argumentation. Indeed, argumentative styles vary according to cultures. It seems that French students are used to a strong thematization of disagreement, and the use of a confrontational style, which does not necessarily lead to an dispute, as illustrated by the above-analyzed hybrid case of Julie, Jérémie and Laurent (cf. 3.1.5). This confrontational style has been attested as typical to the French culture, and is generally opposed to "argumentation by adjustment", which uses a much less polemical tone, and a more implicit expression of disagreement (e. g. Ngo, 2011). The argumentation "by adjustment" is rather an implicit negotiation leading to the evolution of the other's position by successive reformulations. If these two stylistic portraits have been defined on the basis of authentic data, they only correspond to ideal poles that make it possible to situate cultural "tendencies" to argue according to one or the other rhetorical style. Thus, on the confrontation-adjustment axis, the American rhetorical style is closer to the "adjustment" pole than the French style, and can be defined as "to be direct and confront conflict, but in a tactful manner" (Oetzel et al., 2001, p. 240).

These cultural differences may explain the fact that, when students face difficulties to engage into exploratory talk, Americans tend to stick to a consensual identity footing and to fall into cumulative talk, whereas the French adopt a competitive identity footing and fall into a dispute, where personal attacks intensify the confrontational tone in a self-perpetuating system of feeling offended/counterattacking. This is congruent with the fact that no typical case of *cumulative* talk was found in the French corpus and, conversely, that no typical case of dispute was found in the American corpus. Of course, it would be risky to conclude that the French can never engage into *cumulative* talk, or that the Americans never develop *disputes*. These are simply practices that seem to be more marked, scarcer, in these cultural contexts.

3.3.3 Discussion climate: recommendations for practice

Investigating the social aspects of these argumentative interactions, inevitably related to cognitive and emotional issues, sheds light on how and under what conditions constructive group debates may occur. The study of small group discussions, in the whole corpus, shows that it is possible, in different socio-linguistic contexts, to see real sequences of collective exploration of complex and controversial issues. The cognitive skills necessary for such an exercise seem to be well appraised in a transversal way, beyond the linguistic and cultural borders, or those drawn by the different didactic contexts studied. However, it also happens that group debates do not lead to the development of a real in-depth reasoning, when the students face difficulties in managing the relative place of the objects and subjects of the conversation. In order to take in hand a controversial object together, it is necessary to form a group in such a way that each one is serene as to the preservation of his or her face, and dares to show his or her imperfect, under-construction reflection, to the others, with the hope of benefiting from assembling these small pieces of the collective puzzle. In such a configuration, the socio-cognitive conflict appears fully, and can be resolved at the socio-cognitive level. However, sometimes, identity insecurity is such that people are more concerned with saving their face than with the cognitive exploration of the objects under discussion. In this empirical study, it seems that three types of configurations can then emerge: the inability to truly behave as a group, cumulative talk, or the disputational talk. In the first configuration, people do not manage to align their identity footing and to develop a real group talk, and rather develop a hybrid form, which can possibly be transitory and lead, later, to an emblematic type of collective discourse. Another possible scenario is the non-emergence of the controversial part of the cognitive object addressed: we are indeed in a socio-cognitive activity but purely consensual, without conflict, which is limited to cumulative talk. This seems to be the dominant tendency in the American corpus when the students do not manage to engage into exploratory talk. Finally, a third configuration consists, on the contrary, in strongly thematizing the disagreement, by over-investing the object of the debate: it is no longer just a question of choosing and defending rival theses, but of attacking or defending the people who carry them out; the dispute breaks out. In the French corpus, this is the predominant strategy as soon as there is identity insecurity. This leads to focusing on the social dimension of the socio-cognitive conflict, and to resolving it sometimes by social-only strategies of relaxation (making a joke, switching to another topic), to the detriment of a real treatment of the cognitive substance of the problem.

These results invite us to take into account, for practical purposes, four essential aspects that condition engagement into a constructive group debate: the capacity to "form a group"; the management of disagreement; the weight of personal attacks; and familiarity with the notion of controversy. These fundamental elements play a direct role in the overall emotional climate, creating or not a feeling of identity security sufficient to collectively tackle a complex controversy. This chapter points out three directions for educational practice.

The ability to form a group can be worked on through multiple exercises and games, even long before starting the targeted cognitive task. Strictly speaking, it is not an individual skill that each person could acquire or work on independently, even if this ability is based on the social and interactional skills of the group members. It is truly a skill that is built at the level of a given group, which can be strengthened by the time spent together, or weakened by a failure that would question group efficacy (Mullins, Deiglmayr, Spada, 2013).

Managing disagreement is rarely explicitly taught in educational settings. The members of a group in which an attempt is made to create a socio-cognitive conflict are not specially equipped to "manage" disagreement and they improvise a way of doing so, possibly using routines from other social fields. This point should undoubtedly be made more explicit during pedagogical activities based on socio-cognitive conflict, taking the time to collectively establish rules of behaviour, which should not be thought of as universal recipes, but co-constructed as specific principles corresponding to the given situation.

Finally, getting familiar with the notion of controversy is essential for engaging into a constructive group debate on this type of issues, here socio-scientific controversies. This is a matter of epistemic values, bridging ways of discussing and the discussed objects. Indeed, participants should be aware that the problem addressed 1) can only be considered subjectively by nature; 2) conveys diverging interests and 3) questions values, as well as regimes of legitimacy. Such awareness is necessary to overcome the true/false dichotomy, often translated, in a debate situation, by the sterile couple "I am right/you are wrong". Once one understands that viewpoints on such a subject can only be socially situated, and may compete without any of them being contrary to reason, it is easier to respect, even if one is opposed to it, the rival positions and the person(s) holding it; and easier to accept criticism. Thus, it can be very useful, before setting up the debate itself, to work on opposing arguments concerning the same type of controversy, or to compare how different newspapers or social actors deal with it, for example, to favour the understanding of what implies discussing such a controversial object. It can be hypothesized, moreover, that the more a group has worked on controversial objects together, the more it is able to take up a new controversy collectively and constructively.

166

Condemned both by the theory and ordinary argumentative norms, the personal attack strategies are nevertheless commonplace in debates. In this corpus, 4 forms of personal attacks were identified: *ad hominem*, l'*ad personam*, *acccusation of sociocentrism* (suspicion to be led by a hidden mobile corresponding to the interest of a particular social group), *accusation of incompetence*. Surprisingly, a fifth type of personal attack, contradicting the argumentative politeness system, was revealed: *self-depreciation*. Personal attacks are more frequent in typical cases of *dispute*, which is consistent with the interpretation that the feelings of face offense or safety are essential to determine the quality of group argumentation. It would be interesting to work on a different corpus, either of media or political discourses, for instance, to see whether it contains such *self-depreciations* as those that the students produce. This very specific form of self-attack reveals an identity vulnerability, an inextricably social and emotional construct that interferes with collective reasoning. Thus, it seems necessary to pay a special attention to personal attacks at defining ground rules to argue. Presenting the different types of personal attacks and interactively reflecting about their consequences for the target person and the group dynamics might be helpful. I think an open discussion about the relevancy of each ethotic strategy in reference to the problem at stake should not be avoided, given their frequency. Instead of an authoritative moral prohibition to use personal attacks, or of the consensual beautiful vow of always dissociating an idea from the person who defends it, I believe more fruitful to provide the students with reflective tools to get prepared, both individually and collectively to the ethotic strategies that might emerge during the debate.

IV. The importance of objects for argumentation comparative studies

While considering how to foster constructive group debates, one inevitably comes accross the substance of the debate, what is being discussed – either its the controversial nature, or its emotional framing. This is due to the fact that, in practice, it is impossible to fully separate something that would be formal-only argumentative techniques from the cognitive content of the problem at stake. In other words, one does not reason in the absolute but about something, and one never reasons in quite the same way about different objects. At the didactic level, instructional design is not just about training argumentative skills on a theme that would work as a pretext, but to deepen such skills at reflecting upon on a SSC that is fundamental for the future of humanity: the management of drinking water. At the interactional level, taking an emic perspective, i.e. if we look at the recorded conversations from the participants' point of view, their activity is not a training in argumentation, but aims at handling complex socially accute questions, to build and defend an opinion about them. In this last chapter, we will therefore listen to what the students tell us about the substance of the debate, and draw our attention on the conclusions that they reach on the basis of elementary cognitive units, emerging and been challenged during small group work, and finally leading to proposals during the final, whole-class debate. As a reminder, they have to answer the MQ, reproduced below.

> In your opinion, in the future, whether a person has access to dinking water will depend on...?
> a) on how rich the person is
> b) on how physically able the person is to live with lower water quality
> c) on efforts made, starting now, to save water by using less and to protect water resources
> d) on where on the globe the person is born
> e) on nature's capacity to adapt to our needs for water
> f) on scientific advances.

Of course, a great variety of things are said about the debated questions, and I seek here to step back and identify the major trends regarding their content. To do this, I have studied the final whole-class debates of the 10 *cafés* in the *YouTalk* corpus, combining several methodological approaches, the results of which converge to describe some predominant argumentative scripts.

A first method makes it possible to apprehend two dimensions of the way in which the issue is discussed: the thematic anchoring of the debate in one or more fields of knowledge and the analogical cognitive model used to think about the central discourse object that is "water" (3.1). A set of clues elaborated through a convergence of several other methodological tools allows to specify how these discussed objects contribute to argumentative scripts opposing typical arguments and counter-arguments (4.2).

4.1   What are we talking about? Themes and cognitive models discussed

In order to study the content of these final, whole-class debates, a first method of analysis consisted in carrying out a systematic coding of the students' turns of speech, using the *Transana* software, and in comparing the prevaling codes in the different debates. I developed two independent coding schemes to address two dimensions of the debates: the apprehension of the question as belonging to one or more fields of knowledge, or "thematic anchoring"; and the cognitive models used to think about the central object (drinking water), which determines by analogy the types of discourse produced about it. I detail such method illustrating it with the first coding scheme (thematic anchoring), and then present the results of this double coding.

4.1.1 A method to identify the themes of a debate

First, how can we define the thematic anchoring of a debate, i.e. build a global picture of the main fields of knowledge considered at discussing on such issue? The coding of the turns of speech in the video annotation software *Transana* makes it possible to know which temporal frame of each debate corresponds to the discussion of such or such theme, and thus to know which portion of the debate relates to economic or environmental considerations, for example. However, an essential concern of this type of method is to test the reliability of the coding, i.e. the extent to which the segments of discourse that have been coded as "economics" or "environment" can actually be considered as economic or environmental. To this end, I proceeded in six steps.

The first step was to identify the segments of discourse on which the codes would be applied. I chose as the unit of analysis the turn to speach in the public space of the class, that is to say the taking of the floor, most often ratified by the person facilitating the debate, addressing the whole class. It focuses on who holds the main attention of the class group at a given moment. In fact, sometimes there are other, side conversations going on at the same time, without change of the main focus of attention. Thus, several overlapping turns of speech are sometimes grouped together in the same "public turn of speech", as in the following authentic example where 4 turns of speech are attributed to a single "public turn of speech" attributed to Justine:

> JUS   si tu prends la globalité parce que si tu prends au cas par cas euh: [par exemple& (if you take the globality because if you take it case by case uh: [for example&)
> LEO   [oui mais c'est sûr ([yes but it's sure)
> JUS   &[forcément au Cameroun la population elle est (&[necessarily in Cameroon the population is)
> FLO   [xxx

The second step consisted in empirically defining the codes, i.e. the thematic domains considered, through a first exploratory coding. I did not apply *a priori* defined categories but I visualized the videos corresponding to all the different debates, and tried to attribute to each public speech turn one or several thematic orientations corresponding to institutionalized fields of knowledge. Thus, the turns of speech that seemed to me to raise economic issues, for example, were assigned the code "economics". In doing so, I recorded the elements that made me decide to apply this or that category, and then created a dictionary of the different codes used. For example, I noted that the elements that I considered to be sociological in this corpus were mainly related to 4 key points: inequality of access to water according to income, inequality of access to water according to the socio-political characteristics of the countries, the different current and potential uses of water, and what is presented as corresponding to the classic behaviors of people (fashion, mobility, etc.). A total of 11 thematic orientation codes by field of knowledge thus emerged from the analysis of the data: economics, environment, geopolitics, geography, history, metaphysics, physics, politics, health, sociology, technology.

After this exploratory coding, I started coding the debates again, trying to systematize my approach, sticking to the rules defined in the dictionary for assigning codes to public speaking turns. As in the exploratory coding, this second coding proceeded according to the principle of non-exclusivity of codes. Thus, several codes were applied to a single public speech when it referred, fully or in parts, to diverse fields of knowledge.

The fourth step was to get the thematic temporal frame for each debate, i.e., to estimate the proportion of debate time spent addressing the problem from each thematic perspective. This was facilitated by the reporting function of the *Transana* software, which allows us to calculated the global time over which a code is applied. Such temporal frame, expressed as a percentage of the total duration of the debate, gives an idea of the prevalence of a thematic orientation for a given debate. For each debate, I created a radar graph based on these data on thematic frames, showing the relative weight of the eleven different orientations. For example, figure 13 shows the graphs of thematic anchoring for the two analyzed debates hold in the Mexican public school.

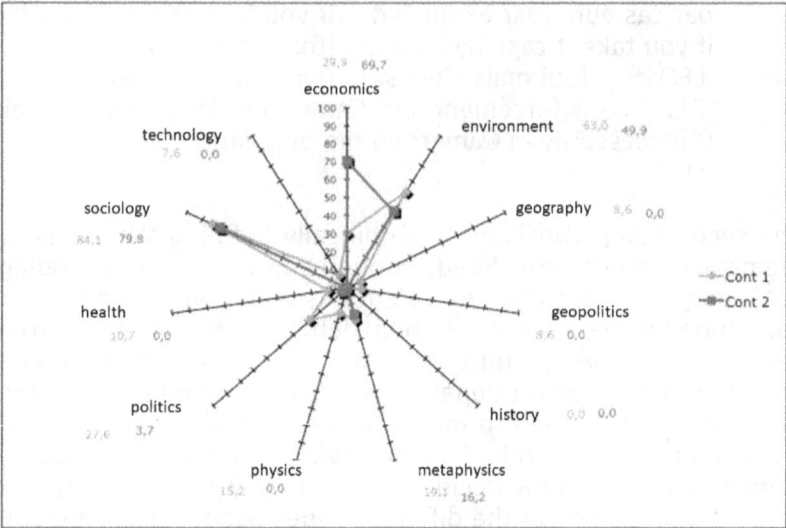

Figure 13. Thematic orientations of the two debates conducted in Contepec, the Mexican public school, in percentage of speaking time.

The fifth step was to validate the coding measuring inter-coder reliability. Two people who were not involved in this research agreed to code, independently, the entire French corpus. They were trained to use the coding guide including the dictionary with examples, during a 5-hour session. In total, 219 extracts were coded independently by the two external coders and myself. Two forms of comparison were then carried out to measure the reliability of my coding, each time by comparing each external coding with mine, as well as by comparing the average of the three analyses with mine, and independently for each code. First, the percentage of agreement was calculated, and then the Kappa coefficient, to measure agreement beyond the effects of chance. With both external coder, the rate of agreement for each code is higher than 80% and 85% for 9 codes (excluding the codes "sociology" and "politics"). This rate is above 90% for the comparison with the average of the three analyses, with the exception of the code "sociology" (88%).

The Kappa values obtained are, with one exception, greater than 0, indicating agreement above that which could be reached by chance. However, they are rarely very high compared to the 0.8 or 0.7

172

threshold often used in publications describing quantitative analysis methods. Regarding a 0.7 threshold, the codes "economics", "environment" and "technology" are valid for both external coders. The comparison with external coder 1 also validates the code "physics" for a threshold of 0.7. Applying this threshold to the comparison of my coding and the average of the three analyses, only "politics" and "health" are slightly below it. But the Kappa values considered acceptable vary across the research communities. At the other extreme, some consider Kappa values above 0.4 to be sufficient. This threshold would validate all the codes, except the one corresponding to "health" (0.23 for external coder 2), and the "history" code, specific to the fact that the calculation is, in this case, applied to a very limited amount of data.

A sixth step was made necessary by such results regarding the reliability of coding: specifying a possible margin of error. Given that for both myself and the two external coders, this was our first experience with this type of methodology, the average of our three analyses can be considered more likely to be a "fair" coding than my own coding. It is therefore possible to calculate a potential margin of error by comparing my results to the results of the "average" coding. As a result, we can consider that the conclusions drawn from my coding can be mitigated according to this margin of error. This margin of error ranges from 0.3% for the code "history" to 5.8% for the code "sociology".

4.1.2 What prevailing themes tell us

Radar graphs representing the different thematic frames, for each of the 10 cafés analyzed, are available in my dissertation (Polo, 2014, pp. 366–370). Here, I simply summarize these results, highlighting the comparative analysis of thematic anchoring across the three national subcorpora. Table 23 summarizes the average results by school and gives an example of a statement produced by a student each time the topic is present on average at least 50% of the debate time. When a knowledge area is never present on average half the time or more, an example from the school where it is most represented is given. Some trends, regarding the thematic orientations addressed during the debates, seem to be shared across the 10 cafés, namely the strong presence of the fields of economics, environment, and sociology. Among the factors that can explain this observation, two hypotheses can be raised. The first is the possible bias related to the presence of an opinion question concerning a very socio-economic problem immediately before the MQ, in the MCQ (OQ3 on the determination of the price of drinking water), which would favour the prominence of these fields of knowledge in the rest of the activity. The second is the more general context of the introduction of the science café activity within the school, often brought by teachers as an environmental education activity, which would influence the students and lead them to consider the problem from this perspective. In addition, it is interesting to note that metaphysical orientations were identified in all debates, but in medium to low proportions. This could be an indication of the success of the pedagogical activity in fostering the encounter of daily life with more institutionalized elements of knowledge–belief, even if the school context implies a preference for the latter.

Table 23. Thematic orientation: average time frame by school and selected examples.

Schools	Mexican public	Mexican private	American	French
economy	50 %	72 %	53 %	47 %
examples	mx pu: *venderla* (selling it) mx pri: *ya después de haberla comprado* (yet once purchased) USA: *your dad works at a waterplant*			
environment	57 %	55 %	70 %	53 %
examples	mx pu: *la estámos contaminando* (we pollute it\) mx pri: *la tienes que cuidar* (you must take care of it) USA: *the polar icecaps that are melting* fr: *si on fait des efforts c'est pour toute la planète* (if we make efforts it is for the whole planet)			
geography	4,5 %	2 %	39 %	24 %
example	USA: *you gonna have water for a really long time (with) the Michigan*			
geopolitics	4,5 %	6 %	33 %	41 %
example	fr: *quand l'humanité va finir par manquer d'eau (...) ce s'ra la loi du plus fort* (when humanity will end up lacking water (...) it will be the law of the strongest)			
history	0 %	0 %	13 %	0 %
example	USA: *back in the 19th century (...) water would be sold for hundred dollars*			
metaphysics	18 %	35 %	28 %	21 %
example	mx pri: *la: naturaleza no se puede cambiar* (nature cannot be changed)			
physics	8 %	21 %	44 %	18 %
example	USA: *the polar icecaps that are melting*			
politics	11 %	57 %	66 %	30 %
examples	mx pri: *tendríamos ya que ver como algo económico (...) para ponerle un límite* (we should see it as something economic (...) to put a limit on it) USA: *the water like issue*			
health	6 %	37 %	26 %	17 %
example	mx pri: *la B (...) se van a estar peleando (...) por conseguirla* (la B [according to the physical ability] (...) they will fight (...) to get it			
sociology	85 %	77 %	75 %	76 %
examples	mx pu: *como somos los mexicanos de responsables (...) vamos a ahorrar agua/* (how reponsible we are the Mexicans (...) we will save water/) mx pri: *la tienes que cuidar* (you must take care of it) USA: *even in third world countries the rich get the water* Fr: *des personnes qui n'arriveront pas à suivre sauf des personnes très riches* (people who will not be able to follow except for very rich people)			
technology	4 %	18 %	59 %	29 %
example	USA: *a more effective way a less exp- expensive way to: get the salt out of (.) salt water*			

In the Mexican corpus, the issue is mainly treated as a socio-economic (the economic can permeate up to 78% of the discussion time, and the sociological 70%) and environmental problem (thematic orientation present 49.3% to 69.9% of the time, counting the maximum margin of error). The weak presence of the geopolitical orientation (less than 10% of the discussion time) can be related to the representation that these students have of Mexico's place on the international scene, perceived as not very powerful. Finally, the weak orientation towards the field of technology (less than 10% of the discussion time, including the margin of error, with the exception of a café in the private school where this thematic frame reaches 35%) is also characteristic of the Mexican corpus. The hypothesis can be made that the technological solution seems relatively inaccessible to these students, due to its cost and the perception of the country as not participating in the technological race with the world leaders. It is quite consistent that this feeling of inaccessibility is less prevalent in the private urban school, where students belong to a socio-economic elite. The orientation of the debate towards politics, on the other hand, shows a wide dispersion: between 2 percent and 75.2 percent, counting the maximum margin of error. There indicates that this is a sensitive issue: either it is hardly mentioned at all, or, when politics is brought to the table, it takes up a lot of space.

Conversely, the American corpus is characterized by the strong presence of a technological orientation, which, depending on the café, accounts for 38 to 76% of the speaking time. The hypothesis can be made that the American scientific and technical culture, rather positivist, as well as the technological and economic power of the country, favors the perception by the students of technology as a possible solution to the problem discussed. The treatment of the political aspect of the problem is also highly represented, permeating between 53% and 75% of the discussion time (margin of error included).

In terms of orientation towards technology, the Lyon corpus is at an intermediate position between the Mexican and American corpora: depending on the debate, it is discussed between 31% and 55% of the time. On the other hand, it is characterized by a fairly strong anchoring in the field of geopolitics, reaching a thematic frame that exceeds 71% in one of the debates.

4.1.3 Metaphorical reasoning: models of water used by the students

I used the same coding method to address another aspect of the framing of the problem, namely the analogies used to think about the object "water" in the students' discourse. In the spirit of the analysis of schematizations proposed by Grize, I consider that they are real "cognitive models", even if I stick, by the nature of my data, to their discursive manifestation: "Natural logic can be defined as the study of the logical–discursive operations that allow the construction and reconstruction of a schematization. The double adjective is there to underline the fact that we are in the presence of thought operations, but only insofar as these are expressed through discursive activities." (Grize, 1996, p. 65).

Taking an interest in the cognitive models of water used by the students, the metaphorical features that they recurrently associate with the object "water", has two advantages. First, it provides a clue, if not a gateway, to the students' common representations, which are likely to promote or hinder the adquisition of knowledge about the object under discussion. Attention to water models also shows how the argumentative use of analogy works, as Plantin explains: "Using a metaphor, the speaker openly seeks the interpretative cooperation of the audience; creating cooperation, metaphor strengthens the importance of prior agreements. (...) Metaphor applies the language of a model, i.e. the Resource domain (the metaphorical term) to an actual situation, the Problematic domain to which belongs the (sometimes missing) metaphorized term." (Plantin, 2018, "Metaphor, Analogy, Model"). Thus, representing water in a certain way produces constraining effects on the alternatives considered regarding the future of water access, argumentatively *orientating* the discourse.

Table 24 summarizes the average results by school and gives an example of a statement produced by a student each time the model is present on average at least 50% of the debate time. When a cognitive model is never present on average half the debate time or more, an example from the school where it is the most represented is given.

Table 24. Cognitive models: average time frames by school and selected examples.

Schools	Mexican public	Mexican private	American	French
Need related to a public service	9 %	46 %	51 %	37 %
examples	USA: *the lower the water quality that would go to your health*			
Commodity	86 %	86 %	57 %	43 %
examples	mx pu: *depende cuanta agua sea\ le vas comprando* (it depends how much water\ you would be buying (him or her)) mx pri: *la tienes que tener guardada* (you must must keep it\) USA: *you could make like a dig out or small stream that could maybe em go to an area that doesn't have water*			
Raw material	64 %	80 %	78 %	31 %
example	mx pu: *en el norte tenemos agua* (in the north we have water) mx pri: *como cualquier recurso como el petróleo* (like any resource like oil) USA: *you gonna have water for a really long time (with) the Michigan*			

176

Product of a process	33 %	46 %	65 %	46 %
example	USA: *get the salt out*			
Renewable resource	2 %	0 %	4 %	13 %
example	fr: *tu contrôles qu'il pleut ou qu'il pleut pas/* (you control that it rains or that it does not rain/)			

From a methodological point of view, the coding of the cognitive models of water used by the students was carried out according to the same six main steps as the coding of the thematic anchoring of the debate. Five coexisting cognitive models were identified in the students' discourse: they consider water as a *raw material*, like gold or oil; as a storable *commodity,* like a water bottle; as an *abundant, cyclically renewable resource*, like the air; as the *product of an industrial process*, like flour; or as a *need related to a public service mission*, like health. I also tested the inter-coder reliability of the coding of the water models. This test was conducted in conjunction with the coding of thematic orientations by knowledge area with the same external coders. The training they received also included familiarization with the water model dictionary, which also operated non-exclusively. Although agreement rates were relatively high (all above 90% between my coding and the average of the 3 analyses), and all Kappa's were above 0, indicating better agreement than what chance would have reached, most of the Kappa index did not exceed 0.65. Therefore, a margin of error was also calculated for the water model codes. It ranges from 0.2% for the *renewable resource* and *raw material* models to 3.3% for the one of the *product of an industrial process*. The results of this coding also led to the creation of radar graphs that allow for a comparison of the different time frames associated with the different cognitive models, for each of the ten debates studied (Polo, 2014, pp. 374–377).

This analysis shows that, with the exception of one debate in the French corpus, the model of water as a *renewable resource* is very rare (it appears in only 3 of the 10 *cafés*, during 3%, 11% and 38% of the discussion time respectively). The most common models are that of the *commodity* (present in at least 73% of the time in the Mexican corpus and occupying between 37% and 52% of the debate time in the French corpus) and the *raw material* (present in at least 63% of the time in the Mexican corpus and at least 72% of the debate time in the American corpus).

The American corpus has the specificity of being more permeated by the model of water as a *product of an industrial* or commercial *process* (which occupies between 49% and 87% of the debate time). This is consistent with the strong presence in this same corpus of an orientation towards technology.

Moreover, the model of water as a *basic need falling under a public service mission*, which was very present in two of the three debates

177

from the French corpus (respectively 44.2% and 61.7% of the time), is not very present in the Mexican corpus. The hypothesis can be made that the lesser development of public services in this country, and especially in the isolated rural area of the public school where we implemented the activity, limits the students in their propensity to consider this type of solutions. For the two debates from this school, this pattern appears 3% and 15% of the time respectively.

The objects discussed by the students can be apprehended and compared throughout the different debates by rigourously coding speech turn transcripts. I have empirically designed two coding schemes on the basis of the data studied, validated by a good inter-coder reliability. Their application on the YouTalk corpus revealed some regularities but also local specificities. First, the coding of the themes of the debates, corresponding to the fields of knowledge discussed, shows a great presence of socioeconomic and environmental perspectives throughout the corpus, and a limited but constant reference to metaphysics. The importance given to technology characterizes the US debates, while Mexican ones show particularly little reference to geopolitics. Second, the analysis of the cognitive models used by the students to metaphorically think the central object 'water' allowed for identifying 5 fundamental images. Drinking water may be apprehended as a need to be satisfied through a public service (image of health), as a commodity (the bottle of water), as a raw material (as gold), as a product of an industrial process (flour), or as a renewable resource (the air). The most frequent models in the corpus are those of commodity and raw material. The image of the product of an industrial process characterizes the US sessions, while a rare reference to the model of a need to be satisfied by a public service is specific to the public Mexican school.

In order to render more explicit the teaching of argumentative competencies and to favor collective reasoning, a simplified version of such analyses might be discussed directly with the participants to the debates. An a posteriori reflexive work could be organized, in which we could show them the main themes discussed and those absent, and have them clarify what the use of each single cognitive model implies in terms of opinion-building. After such an awareness-building reflective phase, another argumentative phase could be design to let them 1) explore new fields of knowledge that they believe to be relevant, or keep to those identified during the first debate, justifying their choice and 2) specify their position and make it more consistent by choosing the appropriate cognitive model to use, even, if necessary, by developing a new one that would embrace and conciliate aspects of water which were only apprehended so far by the combination of several distinct images. Indeed, analogical thinking, embodied both in speech and gesture may bring great cognitive advancement through the emergence of a synthetizing metaphor through collective reasoning (Polo, Lagrange-Lanaspre, 2019).

4.2   Typical argumentative scripts

Identifying, thanks to a robust coding method, the thematic anchoring of the debates in one or several fields of knowledge, and the cognitive models of water that permeate them, already gives important information on the way the issue discussed is considered. Other complementary methodological tools, whose results converge with these first analyses, allow us to specify the typical argumentative script in this corpus. Textometry proves to be very useful to specify the contours given to water as a discursive object. The comparative analysis of the student opinion polls carried out at different stages of the educational sequence also provides indications on the relative frequencies of the different theses and alternatives actually competing in these *cafés*.

4.2.1   The   'words   of   water':   textometrical   apprehension of the debate object[16]

I carried out a textometric study on the discourse object "water" (*water, agua*) in these 10 final debates using the TXM software[17], consisting in listing the vocabulary associated with it. At the theoretical level, the aim is to identify the different "lights" that allow its discursive construction, reconstruction and transformation during the course of the argumentative exchanges. The main hypothesis at stake is that people defending different opinions in these debates will use the term "water" differently. On a macroscopic scale, the question is to know if specific trends emerge, according to schools and/or countries, regarding preference over certain associations to "water" rather than others, which would correspond to prevailing argumentative scripts or arguments.

The method used is based on the automatic analysis of co-occurrences of TXM, i.e. to identify the occurrences of the searched *pivot term* and to classify them according to their immediate contexts (left context: ten previous words; and right context: ten following words). Such an approach embraces the debate as a whole, and does not take into account the boundaries of the speech turns: if the *pivot term* is located at the beginning or at the end of a turn, the end of the previous turn,

---

[16]This section was the subject of a paper and subsequently published in the proceedings of the first European conference on argumentation (Polo et al., 2016a).

[17]TXM is a textometry software developed by a team of the ICAR laboratory, freely available, and for which trainings are regularly proposed: < http://textometrie.ens-lyon.fr/>. I thank Bénedicte Pincemin, Matthieu Decorde and Serge Heiden for their support in my apprehension of this tool.

or the beginning of the following turn, appears, respectively, in its left or right contexts. In order to get the most exhaustive list of associations with the term "water", I proceeded in three steps for each debate. A first search was carried out with the *pivotal term* "water", in the singular and plural, then the occurrences were classified, by left context, then by right context, which made it possible to identify a first list of possible contexts. In a second step, words that could be used as pronouns referring to "water" were searched as *pivotal terms*, one by one. When not previously listed, the results were added to the list of occurrences. If the contexts differed from the previously listed contexts, they were also added to the list of possible contexts. Finally, a last step consisted in carrying out systematic queries using the possible contexts listed during the first two steps as *pivotal terms*, in order to obtain possible other occurrences referring to "water", which had been ignored until then.

The results show similarities between the debates organized in the same country, characteristics that contrast with those of the debates that took place in other countries. Figures 14, 15 and 16 summarize the results of this analysis for each country. They display several aspects of the textometric analysis. First of all, they show the variations in frequency in three ways: each term is associated with a number, which corresponds to its frequency in the corpus; the terms also appear in larger fonts when they are more frequent; the terms with more than 3 occurrences are slightly shifted towards the center, symbolically closer to the word "WATER". The verbs used are reported in the infinitive, except when they appear only in a specific expression. Words of the same morpholexical family (noun and verb; singular and plural, etc.) or with the same meaning (e.g., desalinate or remove salt) are grouped together and the font size chosen corresponds to the sum of the frequencies of all the grouped words. The positioning of the terms in space is intended to reflect the way in which they are typically associated with the object "water", which occupies a central place. In the left half of the figure are listed the elements usually associated in the left context of the word "water", and in the right half, the elements usually placed in the right context. The dedicated expressions including left context and right context are directly attached to the word "WATER" in the middle, and framed. In addition, the words are listed from top to bottom in alphabetical order. Finally, a color code has been used to highlight the 5 major types of "lights" – or perspectives– revealed by this analysis: in blue we find what relates to the natural sources of water; in red appear the characteristics of water allowing more or less to satisfy our needs (either qualitatively or quantitatively); in green are written the terms referring to the possible uses of water, including more or less environmentally–friendly practices; brown is used for elements that reflect the point of view of the drinking water producer, either about production techniques or the problem of supplying consumers; and purple is used for everything that refers to the commercial transaction of which water may be the object. Black remains the "neutral" font used for terms depicting the general issue of access to water, in a more abstract sense.

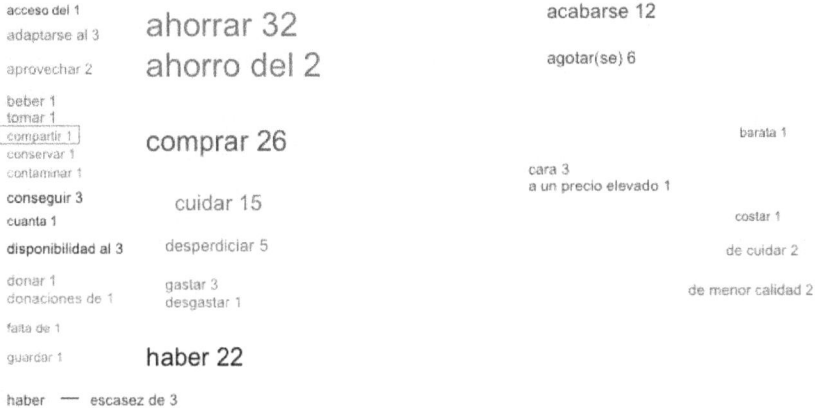

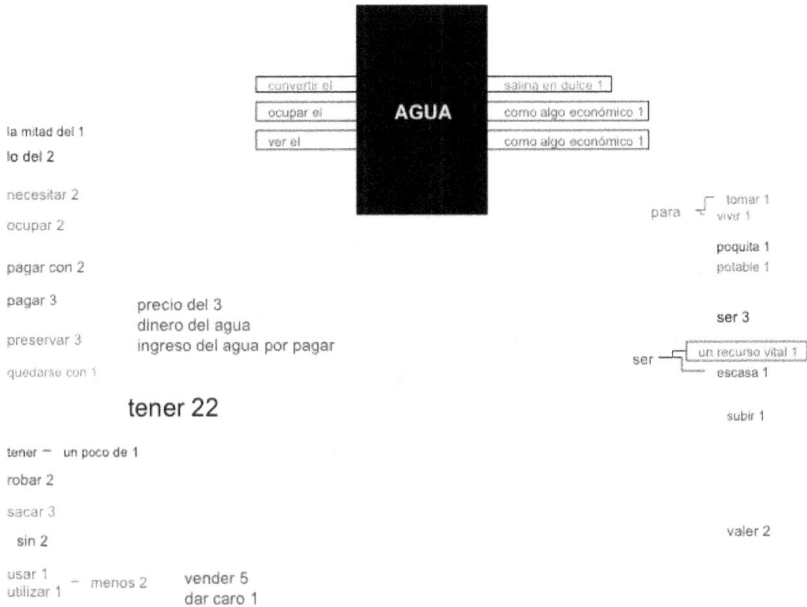

acceso del 1
adaptarse al 3
aprovechar 2
beber 1
tomar 1
compartir 1
conservar 1
contaminar 1
conseguir 3
cuanta 1
disponibilidad al 3
donar 1
donaciones de 1
falta de 1
guardar 1
haber — escasez de 3

ahorrar 32
ahorro del 2

comprar 26

cuidar 15
desperdiciar 5
gastar 3
desgastar 1

haber 22

acabarse 12

agotar(se) 6

barata 1

cara 3
a un precio elevado 1

costar 1

de cuidar 2

de menor calidad 2

la mitad del 1
lo del 2

necesitar 2

ocupar 2

pagar con 2

pagar 3

preservar 3

quedarse con 1

tener 22

tener — un poco de 1
robar 2
sacar 3
sin 2
usar 1
utilizar 1 — menos 2

convertir el
ocupar el
ver el

AGUA

precio del 3
dinero del agua
ingreso del agua por pagar

vender 5
dar caro 1

salina en dulce 1
como algo económico 1
como algo económico 1

para — tomar 1
vivir 1

poquita 1
potable 1

ser 3

ser — un recurso vital 1
escasa 1

subir 1

valer 2

Figure 14. Schematization of 'water' in the Mexican data: visualization of co-occurring terms, their focus and frequency.

In the Mexican corpus, the point of view of the producers or providers of drinking water does not appear much. The debates in Mexican schools mainly deal with water from the consumer's point of view, associating it mainly with the terms *ahorrar* and *cuidar* (to save and take care of), *vital* (vital), *agotándose* (exhausting) and *cara* (expensive). The concerns and practices of consumers are widely detailed, the key issue being to get and use well the water needed to live. The 'light' of the object "water" referring to its ability to satisfy human needs is indeed presented here with a strong emotional component positioning the problem as a question of life and death. Moreover, the frequent mentions of the diminishing resource (18 occurrences of *acabarse*, to end, or *agotarse*, to exhaust) and of the high and increasing price of water depict access to water in terms of socio-economic inequalities. This high cost of water is in line with the recurrent promotion of practices to save and take care of it (37 occurrences of *ahorrar*, to save, or *guardar*, to keep, or *usar/utilizar menos*, to use less, and 9 occurrences condemning practices of pollution – *contaminar* – or waste – *desperdiciar, gastar, desgastar*).

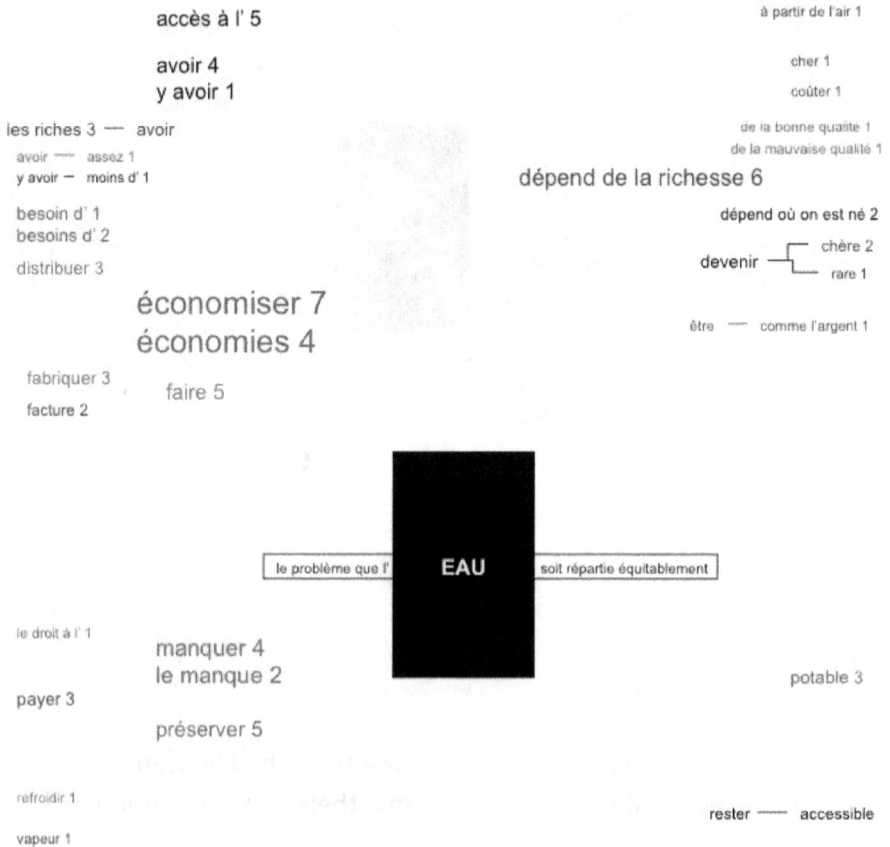

Figure 15. Schematization of 'water' in the French data: visualization of co-occurring terms, their focus and frequency.

Four central terms are used in association with the discourse object "water" in the French debates: *save*, *buy*, *distribute*, *right*. Water seems to be very rarely discussed in terms of its natural sources (only 3 occurrences) and, when it is, it is mainly to say that it is becoming scarcer. What emerges mainly from the discussions is a call for a reasonable use of water: the 16 occurrences concerning the uses of water are all linked to the idea of saving or preserving it, but, paradoxically, no example of a concrete "good practice" appears. The issue of economic inequalities in access to water is also very often mentioned, and is seen as a social problem that needs to be solved by better organization of water distribution, considered as both a need and a fundamental right. Indeed, we note 16 occurrences addressing human needs for water, 12 that depict the problem from the point of view of the water supplier, and 10 associations referring to unequalities in access to water, according to people's income.

access to 10
accessibility 1

affordable 1

abuse 1
available 1
channel the 1
conserve 1
consume 2
economize 1
doing 1
drinking 1

bad 2
the worst of the 1

cheap 3

consumption 1

expensive 1

desalinize 2
desalinization 1
take the salt out of 1

get 13

from the polar icecaps 1
goes1
has been taken 1
is 1
is out to bigger countries 1
issue 1

pay to 1 — get
the riches 1

get — a lot more of 1
that much 1

give 1
give 2 — them 2

good 3
better 1
have 5

as much		as we want 1
give		to other countries 1
have much	WATER	left 1
leave the		on 1
lower the		quality 1
make		available 1
reduce the		that we search 1

have —— enough 1
interchange for 1
lot of 1
ton of 1
lack of 1
make 1
need 1
needed amount of 1

lines 1

plumbing 1
problem 1
purification 1

no 2
out of 1
pay for 1
pollute 1
shampoo on 1
populated with 1
produce 1
provide 1
purified 1
renew 1
reuse 1
sell 2
thin 1
waste 1
wrong with their 1

ocean 2
salt 3
protect 3

run out of 3
run low on 1

save 11
usage of 1
use 3
use 3 — less 3

ressources 1

scarce 1

sources 3
supplies 1
supply 1
supplied 1

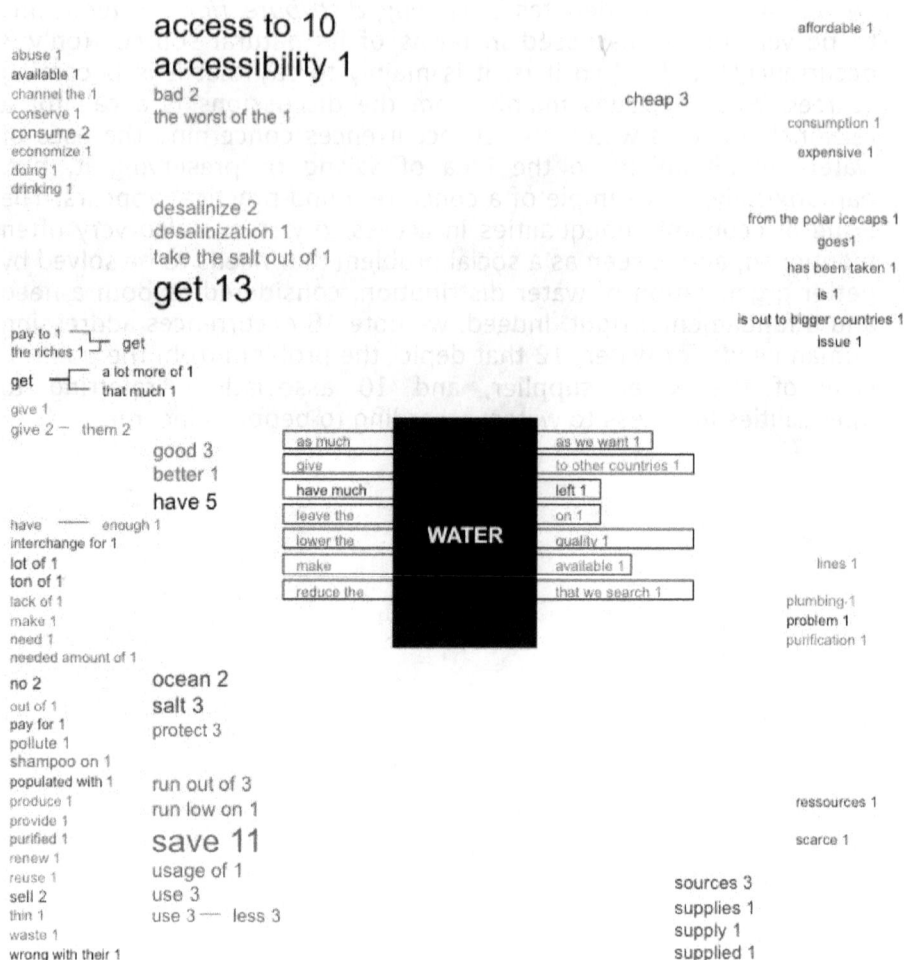

Figure 16. Schematization of 'water' in the US data: visualization of co-occurring terms, their focus and frequency.

On the other hand, in the American corpus, it is rather the point of view of the water supplier that predominates (19 terms), in frequent association with the terms *get, produce, save, good, cheap, sources*. The challenge is to provide quality water at low cost in order to meet the basic needs of consumers, who must nevertheless moderate their excesses. Rather than worrying about water resources, it is a question of technical strategies to make the most of the natural endowment of each territory (particularly developed in this corpus: 23 associations). As for human needs for water, they are evoked, here more than in the debates hold in other countries, from the point of view of the quality of the water supplied (10 associations), and not only through the quantitative concern of having enough water (8 associations). Among the uses of water, beyond the general terms, 20 associations put forward good practices, and 5 habits to be blamed. It is a question of both economizing water and preserving it, two semantic dimensions that already coexist in the use of the verb *to save*. The issue is not viewed from the perspective of the water supplier in the same way as it is in France: what prevails is not the mission of public service but the technical problem of producing and bringing drinking water to the consumers. Consistently, the market transaction to get water is described from both the seller's and the buyer's sides. Unlike in the Mexican data, when the price of water is mentioned, it is rather to present it as relatively cheap. Finally, the American sub-corpus is characterized by two particular elements of language concerning the *schematization* of water as a discourse object. First, the verb *to get*, is used massively, emphasizing more voluntary action than the verbs "to have" or "there is", which are common in the other sub-corpora. Marked on the side of action rather than fatality, this use participates in lowering the emotional tone on the axis of intensity, for the American *cafés*. The second striking element is the designation of the topic by terms describing an abstract issue rather then a concrete problem (*water problem*, *water issue*, *it is wrong with their water*), which also raises the emotional distance, lowering again the emotional tonality.

4.2.2 Opinion surveys and argumentative scripts

Convergences appear among the results of the two coding analyses and those of the textometric analysis: the thematic orientation towards technology in the American corpus, for example, is consistent with the frequent association of water with the technical aspects of its production and distribution, considered from the point of view of the drinking water supplier. Another method produces results that make it possible to identify the argumentative scripts embodied in these debates: the precise analysis of opinion polls about the MQ. This involves tracking the anonymous individual positioning at the beginning of the session, the public choices made by groups of students after the discussion time in 3 or 4 students, and the final anonymous individual vote after the final whole-class debates (the ones that were coded and submitted to the textometric analysis).

• *Typical oppositions: competing alternatives in each school*
The analysis of these votes shows the predominance, in the Mexican *cafés*, of the opposition between answers A (access to water will depend on how riche you are) and C (it will depend on the efforts we make). However, differences appear between the two Mexican schools. In the rural public school, the responses the most initially chosen are

C and D (access to water will depend on where you were born), with C being about twice as chosen as D (49% versus 23% on average). Analysis of individual voting trajectories, which informs us about opinion shifts from one option to another from the beginning to the end of the café, reveals that most shifts were made to options A (involving an average of 5% of initial choices and 21% of final choices) and C (average of 56% of final opinions). In the urban private school, option C is the most chosen initial option (55% on average), followed by A (19% on average). The transfer study indicates that students who change their opinions during the *cafés* also move primarily toward these two options. Some students switch from C to A over the course of the activity, but option C remains the most chosen, especially by students who initially chose option E (access to water will depend on nature's ability to adapt to our needs). Finally, option F (access to water will depend on scientific advances) is very rare in the entire Mexican corpus.

These results are consistent with those of the thematic and cognitive models analyses. Indeed, the sociological anchoring is the prevailing one in these debates, and it is associated with the choice of options A and C, which, respectively, also include a reference to the economic and environmental fields. These two options do not refer to the model of water as a cyclically renewable resource, nor to the model of a basic need associated with a mission of public service, both of which are very rare in the Mexican corpus. On the other hand, both relate to water as a storable commodity, the dominant model in these *cafés*. While response A can potentially be associated with the model of water as a product of an industrial process, and response C with that of water as a raw material, the image of the raw material is much more present in this corpus than that of the product of a process. The results of the textometric study indicate that the script opposing A vs. C, in the Mexican corpus, is rather based on the point of view of the water consumer, and very little on the one of the water producer, which is consistent with an under-representation of the model of water as product of an industrial process.

In the U.S. data, options C and F also dominates (chosen by an average of 31% and 37% of students respectively at the end of the session). Options B (access to water will depend on the physical ability to live with lower quality water) and E (access to water will depend on nature's ability to adapt to our needs) are little chosen in the initial vote and very little chosen thereafter. The most frequent initial opinions are C (49% of the votes on average) and D (access to water will depend on where we are born, 25% of the votes on average). Option F, which averages no more than 15% of initial votes, gains weight over the course of the debate, and is very frequently chosen in the group and final votes.

These results also converge with those of the coding analyses: the importance of the theme "environment" is linked to the discussion of options C and F; the medium place of "geography" and "geopolitics" is linked to the initial weight of option D; the frequency of the theme "sociology" is linked to the importance of option C. On the other hand, the strong presence of the economic dimension in these *cafés* contrasts with the fact that it is not directly linked to response options C, D and F. This economic coloration also appears in the results of the textometric study, with a set of associations relating to the commercial transaction of which water may be the object. The technological dimension associated with response F is also quite frequent, a trend confirmed by the textometric study, which highlights a strong focus on the debate from the point of view of the water supplier, in its technical aspects. Although options C, D and F theoretically allow all the cognitive models of water to be covered, with the exception of the renewable resource model, which is not very frequent in the corpus, these models are unequally mobilized. The models of raw material and product of an industrial process are over-used in comparison with those of storable commodity and fundamental need associated to a mission of public service. This observation converges here again with the strong focus on the perspective of the drinking water producer, as identified by the textometric indices.

In the French corpus, the rivalry rather opposes options A or D to option C (corresponding on average to 38%, 24% and 28% of the votes respectively). Options B (access to water will depend on the physical capacity to live with lower quality water) and E (access to water will depend on the capacity of nature to adapt to our needs) are rarely chosen in the initial vote, never chosen in the collective vote, and rarely chosen in the final vote. Option F is more frequent than in the Mexican corpus, and much less than in the American corpus. The opinions mostly chosen at the beginning of the *café* are also mainly C and D (gathering respectively 39% and 33% of the votes on average). Option A appears during the debate, receiving most of the votes from students who had initially voted for C or D.

The predominance of the sociological dimension of the problem, identified through thematic coding, is consistent with this finding, since it is linked to responses A and C. Similarly, the environmental and economic dimensions are respectively associated with options A and C, and correspond to the two other largest thematic orientations of the French corpus. The fields of geography and geopolitics, linked to option D, occupy a medium place in these debates. The technological dimension also has an limited scope, a weight consistent with the results of the textometric study, which reveals an medium presence of water schematization from the point of view of its producer or provider. This observation converges with the relatively strong presence of the model of water as a fundamental need falling under a mission of public service, in this corpus, even though this code is not obviously associated with any of the options A, C and D. In the Lyon context, the argumentative opposition script that pits them against each other is thus marked by a political orientation, in the sense of public policy. Discussing these three options could, conversely, have led to the mobilization of the cognitive model of raw material, which is relatively little represented. The product of a process model seems to be rather over-represented in relation to the alternatives discussed. This last point should be interpreted in relation to the fairly strong presence of water from the perspective of its commercialization, as identified by the textometric analysis, and which corresponds to one of the defining features of the codes "product of a process" and "storable commodity".

- *The agenda-defining function of group vote*

The analysis of the opinion polls also reveals that, in all the schools, the pedagogical setting had an effect of polarizing opinions around a few response options, two of which often encompassed 80% of the final votes. This polarization effect should be interpreted regarding the collective votes cast during the *cafés*. Indeed, the latter seem to determine in a crucial way the framing of the final debate in the whole class, the focus on certain response options, and, therefore, certain aspects of the problem discussed. Thus, the two most chosen options in the final vote, for each *café*, were chosen by at least two groups during the collective vote (with one exception). The collective vote seems to have a function of agenda-defining for some options that, being more discussed more during the whole-class debate, receive more individual votes at the end of the session.

- *Fatalism versus different forms of voluntarism*

Let's quickly go back to the content of the debates revealed by the analysis of the opinion polls. While answers A and D embody a fatalistic position, options C and F constitute different voluntarist alternatives imagining ways to change the situation, through efforts (Mexican corpus), and/or through scientific and technological progress (American corpus). The fatalistic position (A and D combined) is predominant in the French corpus, even if option C appears to be a sufficiently credible alternative to garner a good part of the votes. Option F, corresponding to a voluntarist perspective of another kind, based on scientific and technological progress, constitutes in the French data the second alternative to the fatalistic position, which is much less chosen. In the Mexican corpus, fatalistic and voluntarist positions clash according to a pattern in which fatality corresponds more to socio-economic status (A) than to place of birth (D), while the only voluntarist alternative is based on daily efforts to save water (C), and certainly not on faith in science (F). Conversely, in the American corpus, the fatalistic position corresponds more to the place of birth (D), and is in the minority compared to the voluntarist position (F and C combined), the preferred alternative corresponding to scientific progress (F).

4.2.3 Comparative argumentation studies: script preference

Although multiple variables come into play in the course of each of these debates, which are not experimental but ecological situations, this international corpus makes it possible to build a contrastive study of students' argumentative practices in different linguistic, cultural, socio-economic and educational contexts. Unlike research on intercultural communication, which focuses on the study of rhetorical styles and/or dominant genres according to cultural contexts, comparative argumentation studies must, in my opinion, take the substance of the exchanges into account. It is a matter of identifying the substratum of informal reasoning which, even if deeply shaped by culture and language, cannot be fully apprehended by a normative framework specifying the forms of a "good" argument. Indeed, such a normative framework is by definition context-dependent (cultural, institutional, etc.) and tends to make invisible all reasoning that deviates from it, recognizing logic only when embodied according to the ongoing norms. This is, for example, the fallacious observation that indigenous peoples in North America are not capable of arguing because they do not produce a structured argument like that of the colonists but merely "give examples," while arguing from an example is a form of reasoning that has been well recognized, including in Western culture, for centuries (Plantin, 2018, "Example"). Such an approach to comparative argumentation studies, as any fine contrastive analysis, implies suspending judgment and not hypothesizing differences or similarities between the three national corpora, but seeking to describe the contours of the predominant argumentative scripts in the exchanges within each of these corpora, and then being more or less surprised by their similarities or differences. Grize's informal logic (1990, 1996) appears particularly valuable for this purpose, as it is based on the *discourse objects* themselves, entities studied in their dual linguistic and cognitive nature. The three methods described above thus aim to qualify the argumentative framing of the debates, by tracing the (re)constructions and focuses conferred by the students on the issue under debate. What differences can we observe between the discourses of students from different countries in such framing of the debate? To answer these research questions, I used the three methods described in this section, which it is time now to triangulate.

Thus, their results converge towards the description of the same phenomenon from different indicators, namely the emergence of a predominant argumentative script in each corpus, which governs the sets of arguments *vs.* counter-arguments that appear most frequently in the data. Finding differences among the national sub-corpora is unsurprising, and can be explained by the prevalence of pre-existing *interdiscourse* (Amossy, 2006, pp. 94–99). For the American subcorpus, the following characteristics were identified:

1. a strong orientation towards the field of technology;
2. extensive use of the water model as a product of an industrial or commercial process;
3. a high frequency of words associated with "water" that apprehend it, as a *discourse object*, from the point of view of the producer of drinking water, particularly the technical vocabulary concerning potabilization processes;
4. a tendency to oppose option C ("future access to water will depend on efforts to save it") and option F ("future access to water will depend on scientific advances").

These four types of results converge to describe different elements of a typical argumentative script that dominates discussions on drinking water in the American school. The problem here is rather technical, finding the most efficient ways to manage water resources, whether it is to optimize their use to reduce water consumption or to consider new techniques to produce more water or to better depollute water. Such a script covers only part of the "argumentative script" attached to the issue, understood as "the set of standard arguments and rebuttals mobilized by one side or the other when the issue is debated" that "pre-exists and informs the concrete argumentative discourses, of which it constitutes a determining, but not unique, element" (Plantin, 2018, "Argumentative Script"). The comparative study conducted, by each of these methods, shows that it is other elements of the "argumentative script" that Mexican and French students pay the most attention to.

Thus, the Mexican corpus is characterized by:

1. a relatively weak anchoring in the knowledge areas of technology and geopolitics;
2. the scarcity of the water model as a fundamental need associated with a mission of public service, and the predominance of the raw material and commodity models;
3. the high frequency of words associated with "water", apprehending it from the consumer's point of view, with an emphasis on the concern to use well the diminishing resource,

necessary for basic needs, but which access is restricted by socio-economic unequalities;

4. a tendency to oppose option A (access to water will depend on how rich you are) with option C (access to water will depend on efforts made).

The problem is therefore clearly identified as a social issue, but also, fundamentally, as a very concrete individual concern, the acute awareness of which is linked either to personal experience of the difficulty of accessing quality water, or to having already witnessed such situations. Water is a scarce but vital resource and is therefore considered an expensive commodity by definition. Little hope relies in the collective organization of its distribution. The most optimistic are betting on the transformation of individual practices in order to save the resource, while the pessimists predict the persistence or even the reinforcement of socio-economic unequalities reagarding access to drinking water.

As for the French corpus, it has the following specificities:

1. a relatively strong anchoring in the field of geopolitics;
2. the frequency of water models of a commodity or basic need associated with a mission of public service;
3. a tendency to associate water with terms that shed light on the socio-economic unequalities that govern its access, and the problem of its distribution and proper use as a social problem, both a vital need and a fundamental right;
4. the typical opposition between fatalistic options A (access to water will depend on how rich you are) or D (access to water will depend on where you were born) and option C (access to water will depend on efforts made now).

Although the opposition between a fatalistic pessimism and an optimistic perspective based on a reasoned use of water can be found in these data, as in the Mexican corpus, this script places more hope in the implementation of a collective response to the problem. The vital need associated with access to drinking water is translated into legal terms, as a fundamental right, for which public power must be accountable, on a national but also international scale.

Textometrics, and, more especifically, the study of the terms associated to the discourse object 'water' reveals several different lights cast on it during these debates. 'Water' may be apprehended from the perspective of its natural sources ; focusing on the way it satisfies human needs ; through the scope of its possible uses ; as resulting from human production and distribution ; emphasizing the payment transaction it may be involved in; or even just as a general social issue. Here again, even if some trends recur in the data, some differences appear between the national corpora, and even sometime between the two Mexican schools. Such textometrical analysis converges with the results got through coding themes and cognitive models in some points. Eventually, the global picture is completed by a quantitative study of opinion surveys all along the YouTalk. It reveals transfers from one alternative to another; and the agenda-putting function of group vote. At the end of the day, each national corpus can be characterized by a specific argumentative script that prevails in the debates, even if some other scripts, in minority, may also emerge. Such results show the interest of multi-methodological approaches, making it possible to replicate similar findings with different, though converging types of evidence: "replicability of results with different methods and types of evidence is actually the best measure of their robustness" (Arppe, 2010 : 6). On the research plane, such a triangulation illustrates the possibility of argumentation comparative studies that, beyond rhethorical styles, could describe debate content-related specificities in different cultural areas. This type of study was undertaken here independently from the work on the descriptive, prescriptive and affective logics, to make analyses easier, since it is always difficult to apprehend all at once the complexity of reality. Still, such study on the objects discussed during the debates and how a specific light is cast on them is perfectly consistent with the tridimensional approach of that I defend, as an activity both and inextricably cognitive, social and emotional. Indeed, the literature has already shown long ago that individual logico-discursive practices are structured by socially and historically grounded forms of discourse, their present argument actualizing typical scripts of opposition in a unique way. In particular, such actualization depends on 1) what the locutors believe to be contextually relevant, according to their understanding of the discussion as a social situation, and 2) the affective framing of the problem, which itself is strongly related to their cognitive investment.

Conclusion: Constructive exploration and socioaffective argumentation

Which lessons can be drawn from the research presented here for the constructive exploration of SSC? From an epistemological and theoretical point of view, it seems important to me to underline the conceptual contributions made in an inductive linguistic analysis approach, going from the data to the model, allowing to finely characterize what is a "constructive exploration" of a SSC. I then recall how this conceptual work can inspire concrete tools for apprehending authentic practices of SSC debate, whether they be research methods or pedagogical recommendations. Finally, I sketch out avenues for future discussion and research on the implications of conceiving the didactization of controversies as a discursive activity consisting in reasoning together in emotion.

From deep attention to data to theoretical interpretation

Thanks to extensive fieldwork, which enabled the constitution of a large audiovisual corpus collected in an ecological context, I undertook an in-depth argumentative study of discussion practices around SSC on water management, in 4 schools in 3 countries. Trying to account for and elicit the abundant discourses produced by the students through the prism of argumentation, new analytical categories emerged, filling gaps in the literature and leading to theoretical proposals for describing daily reasoning.

Thus, distinguishing descriptive, prescriptive and affective logics, in order to better understand how they jointly participate in the construction of an argument, shows the argumentative value of statements that do not have the formal structure theoretically associated with the production of arguments. This makes it possible to highlight the exploitation, in argumentative discourse, of all the resources that participate in reasoning, including as elementary cognitive units affects, norms and values as well as beliefs or knowledge. The proposed typology of the elements of knowledge-belief used by the students also sheds light on the logical and epistemic differences between the various statements presented in these discussions as undiscussed facts. This grid thus clarifies what the frequent assertion that SSC convey "hybrid knowledge" means (cf. 2.2.2). Moreover, the adaptation of the distinction *thymical/phasic* to the analysis of debates, the *thymical* referring to the background emotional tonality and the *phasic* to the emotional variations associated with defending competing conclusions, makes it possible to think of the *affective logic* beyond the individual subject, and to apprehend the argumentative interaction itself as a place of emotion. Thinking about *affective logic* at the collective level favors the understanding of the role of emotions in argumentation: it highlights the consubstantiality of the defense of an argumentative conclusion with the adoption of an adequate emotional position, embedded in the discourse of people sharing the same opinion, who, *align* themselves respectively. This conceptual change allowed us to model the cognitive and social functions of emotions in group reasoning and to generate hypotheses on their relations (see 3.2). Deepenning the understanding of engagement into constructive group talk by considering it as the result of an *alignment of* the individual *identity footing* on a specific face-preservation system constitutes an significant theoretical proposal to complete the model and t reflect on the articulation between the individual and the collective levels. The identification of the four refutation strategies typically associated with the use of general principles (cf. 2.3.3) also proceeds from an inductive modeling, starting from the data to inform the theoretical interpretation of the observed phenomenon. To make sense of the redundancy of general principles in this corpus, I drew on the conversational structure of ordinary argumentation depicted by Muntigl and Turnbull (1998) to derive a model of the exploitation of general principles in argumentative interaction. Thus, when a speaker mobilizes a general principle g to support an argument or counter-argument, four typical refutating strategies can follow, alone or in conjunction:

1. accepting the principle but challenging the conformity of the argument to the principle;
2. accepting the principle but questioning its contextual relevance;
3. rejecting the principle;
4. opposing to this principle by the appealing to an alternative one, a g' principle presented as superior.

This last strategy, based on a rival general principle, allows us to explain how argumentative discourses are born and structured in the form of scripts where typical arguments are opposed to each other, shaping networks of principles against which the different propositions and counter-propositions are evaluated. The comparative analyses carried out on the logico-discursive objects constructed (cf. 4) testify to the existence of such scripts in the YouTalk corpus, and allow us to identify some predominant scripts associated with the different national sub-corpora (cf. 4.2.3).

Concepts to describe and tools to act

Beyond such theoretical contributions embodied in a meaningful analytical language, sometimes going as far as modeling the observed phenomena, the work that I have reported througout this book provides concrete tools that are meant to be used in other contexts, of two kinds. First of all, the strong empirical dimension of this research made it necessary to find adequate concepts to describe the observed practices, which has led to the construction or refinement of analysis techniques "operationalizing" these concepts for discourse analysis. Thus, several methodological tools can be reused in other research. For example, this type of analyses can apply to philosophical discussions, a reseach object around which an interdisciplinary international community is being structured, at the crossroads of linguistics and pedagogy (cf. for example Simon & Tozzi, 2017). Based on Plantin's (2011) seminal work applied to SSC discussion, I have highlighted how the affective tonality of a discourse object, which is to say the cognitive function of emotions in argumentation, can be characterized on the basis of 7 parameters, participating in two fundamental axes: valency (pleasure–displeasure) and intensity (strong emotion – weak emotion) (cf. 2.4). Distance to the problem, possibility of control, description of a causality and/or an agency characterize the emotional *schematization* on the axis of the intensity. The anticipated consequences, the use of more or less positive analogies, position on the life–death axis and (non-)conformity to norms determine the emotional *schematization* on the valency axis. These parameters constitute compasses for any linguistic analysis of the discursive construction of emotions, in the whole spectrum of semiotization, accounting for more or less implicit expressions.

On the plane of group argumentation, I have operationalized the concept of "exploratory talk" (Mercer, 1996) by specifying it into five indicators of the quality of group talk (cf. 2.1):

1. presence of justifications;
2. topical alignment;
3. critical review of all proposals;
4. collaborative decision-making;
5. dialogic strenghtening of arguments.

This last indicator is particularly interesting for determining, in complex cases, whether or not a dialogue with hybrid characteristics with respect to the first four indicators, can be considered as *exploratory talk*. Indeed, by focusing the analysis on the substance, i.e. the circulation of arguments, this indicator allows to avoid the pitfall of basing group talk analysis on only formal, culture-dependent criteria. Moreover, by proposing to multiply the scales of analysis of the types of group talk while respecting the sequential nature of verbal interactions, the linguistic approach presented allows for a better understanding of atypical extracts. Working at a finer grain unveils the specific functions performed by the three types of talk (*exploratory*, *cumulative*, and *disputational*), especially in opening or closing (sub)sequences. It also makes it possible to specify the nature of hybrid transitional sequences during which the members of a group gradually *align* their self-identity footing.

Finally, at a macroscopic level, I carried out a work to identify typical argumentative scripts in each national corpus, providing innovative methods for comparative argumentation studies. Thus, the use of textometry on oral transcripts to characterize the *schematization* of a discourse object at the heart of the controversy (cf. 4.2.1), or the systematic coding of the content of argumentative utterances regarding thematic anchoring and cognitive models (cf. 4.1) constitute operational tools for comparing the substance of the debates, methods that can be used for other corpora.

The concern for going beyong theoretical concepts, towards designing tools also includes a real consideration of the pedagogical objective of fostering constructive debate on SSC. These results argue for taking into account the socioaffective aspects of such debates, from a priori instructional design, to real-time orchestration of the pedagogigal sequence (Dillenbourg & Jermann, 2010).

One of the factors that make it possible to feel safe in terms of face-preservation, an attitude associated with engagement into collective exploration of a controversy, is the ability to form a group, which can be worked on in the long run, through a variety of activities. It is a matter of regularly training the social and interactional skills of individuals in the service of collective cognitive tasks within the same group, in order to build a positive common experience that gives confidence in the ability to work well together (or "group efficacy" described by Mullins, Deiglmayr, Spada, 2013).

At the very moment of setting up such dialogical teaching tasks, the clarification of rules on how to work in groups appears of crucial interest. Educators should foster collective reasoning by making the use of *exploratory talk* THE socially relevant communicative genre in the pedagogical situation: a place where any idea can be formulated, where all options are examined without danger for the *faces of* the people bringing them, where doubt is conferred a great value, and the participation of all group members is encouraged. Two points deserve to be explained in particular: the management of disagreement and the place of personal questioning. Without clear instructions, the members of a group in which an attempt is made to create a socio-cognitive conflict improvise more or less successful ways of managing disagreement: deciding by voting, posting several answers, silencing those who do not think like oneself, discrediting them by attacking them personally, switching to off-task discussions to preserve interpersonal relations... To encourage the most constructive practices for managing disagreement, it may be interesting to establish guidelines at the beginning of the activity, without imposing them as universal recipes, but rather co-constructing common principles appropriate to the specific situation. In this regard, personal attacks, frequent in controversial discussions, deserve a particular attention, because they may lead to a *dispute*, the discussants being stuck in repeating unconstructive exchanges. Explicit anticipatory reflection with the students on the different forms of personal challenging, on the possibly resulting feeling of offence and on their deleterious effects on the group dynamics, and on the relevance of the argumentative use of *ethos* for addressing the substance of the problem can limit the occurrences of unfruitful personal attacks.

In addition to reminding students of these rules, the teaching situation itself can be scripted to ensure that they are followed. For example, the group could be provided with a precise debate procedure implying the successive discussion of options A to F, with each student having to give his or her opinion on each of them before moving on to the next. The hypothesis of a relationship between group talk and thymical framing of the debate also points to affective scripting, in order to promote an optimal emotional intensity and associated level of investment for collective reasoning. This could be done through the use of role-playing games that condition the emotional distance to the problem, or through non-linear interactive narration that allows an event to be experienced through multiple points of view (Polo, forthcoming). In philosophical discussions, the frequent use of examples drawn from personal experience is an element of the pedagogical script that sets a relatively emotionally intense framework for debate, generating investment (Polo, 2020). In any case, this theoretical interpretation condemns a restrictive vision of the problem of emotion regulation, which sometimes is applied as an attempt to isolate "pure" cognitive processes considered as positive for thinking from emotions, caricatured as harmful to reasoning. A more fruitful approach undoubtedly lies in providing the group with tools for awareness and collective regulation. However, emotion awareness alone is not enough to guide students towards cognitive progress. Indeed, students in the midst of an argument may demonstrate a very good awareness of the emotional "heating" of their conflict, but choose social-only resolution strategies based on a cognitive disinvestment (making a joke, switching to another topic, cf. 3.2.3). The most interesting way of taking into account the socio-affective dimensions of reasoning is undoubtedly to construct an initial pedagogical situation that really takes them into account, together with its cognitive dimension. It is a matter of choosing appropriate discussion topics and contextualizing the target knowledge in such a way that supports the right emotional investment of the participants in the group argumentative task. In this respect, the participants need to be familiar with the notion of SSC to engage into a constructive debate. Indeed, knowing that the issue under discussion can only be considered subjectively, conveying interest groups, questioning values, and regimes of legitimacy ; helps to get out of the dynamics of binary opposition. Indeed, once one understands that several points of view can compete without any of them being contrary to reason, it is easier to respect other opinions and the people who defend them. Such familiarization with controversies can be achieved, for example, though pedagogical steps previous to the debate itself, for instance by working on opposing arguments about other SSC, or by comparing the discourses hold on the discussed issue in different newspapers or by several social actors. All these elements my affect the overall discussion climate, aiming at creating the conditions for a sufficient individual feeling of identity safety to apprehend, as a group, a complex problem, embedded in emotions.

Affective reasoning together

The study of this corpus shows that it is possible, in very different contexts, to succeed in exploring SSC collectively. The empirical work carried out also demonstrates that the interactions through which such collective reasoning is built are impregnated with affects: both emotions linked to the discussed objects and emotions linked to the social recognition and affiliation of the debating subjects. Thus, the pedagogical work on socially acute questions such as SSC should consider the socio-affective dimensions of these interactions and the arguments that emerge from them: what is at stake is affective reasoning together. Taking such a concern seriously, and thus overcoming the traditional emotion/reason dichotomy, offers food for discussion and opens up many avenues for future research.

In line with empirical work on the role of emotions in argumentation, the study of affective logic does not try to describe the emotions felt by students, but focuses on the emotions build in their argumentative discourse. Such use of emotions as argumentative resources employs more or less explicit modes of semiotization: shown emotions, said emotions, induced emotions, argued emotions. Using generic emotional parameters, argumentators and analysts can unveil induced, implicit emotions. Actually, while describing the problem, the students produce discursive images, or schematizations of alternative answers (Grize, 1990, 1996), to which they confer more or less negative emotional tones, thus orienting their contribution towards an argumentative conclusion. Since there is no longer any reason to try to chase away emotion, how can we take advantage of such results to inform the teaching of argumentation about SSC? From an explicit teaching perspective, one can argue for a work consisting in making affective logic visible, and fostering awareness of its role in such discussions. Such a position is in line with the concerns of the instructional designers of emotion awareness tools. However, we should remain cautious here: as soon as we understand argumentative situations from an emic point of view, we can hypothesize that if the participants use implicit semiotizations of emotions, it is because it is necessary for them to deliver their full argumentative power. This may be explained by the reminiscence of the ordinary argumentative norm considering affects as obstacles to rational thinking. Overtly using affective logic would expose to accusations of irrational and fallacious reasoning. In this context, pointing out that the supporters of a thesis appeal to a high-intensity emotional framing of the problem may result in immediately discrediting this thesis and its supporters. Any work of emotion-awareness in the argumentation about SSC must therefore be accompanied by previous epistemological training questioning the prejudices about the relation emotion-reason. These precautions taken, it seems to me that it would be interesting to reflect, after the discussion, on the emotionally charged arguments actually produced by the students, and to take the time to make explicit the inferences that underlie them, that is to say to guide the students to pass from induced emotions to argued emotions. Such metadiscursive exercise consisting in arguing a posteriori the implicit emotions exploited in the debate, by bringing out the reasons behind them, would also be an opportunity to highlight the interweaving of affective logic with descriptive and prescriptive logic: the competing emotional schematizations identified may, for example, reveal axiological conflicts (Polo et al. , 2013b).

From a methodological point of view, the use of textometry to analyze the schematization of discourse objects, and particularly its emotional side, constitutes an rich avenue for future work. Indeed, the visualizations produced in the form of word clouds characterizing the apprehension of the issue of drinking water management in each of the national corpora reveal terms with diverse, more or less intense, emotional valencies. For example, positioning oneself more or less as a consumer or producer of water corresponds to constructing a certain emotional distance to the problem. Similarly, evoking water in relation to human needs in terms of degree of quality or risk of shortage leads to considering in the lens of more or less positive consequences, on the emotional plane. Such results can be deepened by using textometry to study the thymical tonality in each subcorpus as a whole. In the work presented here (cf. 4.2.1), the textometric analysis focuses on the debate as a single interactional unit, without distinguishing between speakers: the aim is to characterize the discourse object "water" as it is collectively constructed in the course of the exchanges. One could also imagine, while keeping as object of study collective reasoning about the same problem, to decline such textometrical tools at a smaller scale, on the set of contributions of all the persons defending a common position. Such a study could make it possible to characterize the differences between rival sides in their schematization of the question, notably on the emotional plane, on the basis of the frequency of occurrences of some constellations of words.

Moreover, the project of studying how to reach fruitful affective reasoning together would benefit from really apprehending the interactions in their multimodal nature, including, in addition to speech, written materials, manipulation of objects, proxemics, gazes, and gestural communication. Indeed, recognizing that reasoning about SSC includes dramatical socio-affective dimensions challenges traditional individualistic cognitivism. It is rather consistent with "4E" cognition approaches (e.g. Menary, 2010), which consider learning and reasoning as phenomena that go beyond the limits of the thinking subject (extended), that depend on the environment (embedded), that invest whole bodies and not only brains (embodied), and that are shaped by own the dynamics of living organisms (enacted). Thus, it would be interesting to study how the metaphorical process that supports argumentation by analogy, an key aspect of the use of cognitive models, is collectively constructed in the course of the exchanges, through the use of a variety of communication modalities, in particular thanks to gesture studies. Such a linguistic approach, based mainly on the complementarity between speech and representational gestures, proved useful to follow the construction of collective reasoning in a philosophical community of inquiry (Polo, Lagrange-Lanaspre, 2019). Implementing such a research method for discussions on the "hot" objects that are SSC, by questioning the use of affective logic in metaphorical reasoning, can undoubtedly provide a better understanding of how to fruitfully debate about them. Indeed, the use of metaphor is a great tool to get more familiar, emotionally closer, to an unknown problem, in such a way that the reduced emotional distance does not directly stages the discussants. As such, it undoubtedly plays an important role in the possibility of engaging collectively in the *exploration* of complex issues without falling into a *dispute*.

In terms of recommendations for pedagogical practice, the observation of the agenda-setting function of group vote in these scientific cafés (cf. 4.2.2) leads us to emphasize that the distribution of speaking time in such debates does matter a lot. Indeed, the expressing or not a view in the "public" space of debate (the "class-group"), even non-verbally (by the simple display of a card), has an impact on the range of proposals discussed and selected for the construction of opinions. It seems all the more important to equip these situations with means (either verbal and/or non-verbal) that favor the expression of a diversity of points of view, including minority ones. This is in line with the fundamental rule of encouraging the participation of everyone in the discussion: a controversy is best explored "together" if all the members of the group have a say in it. In this respect, the design of collaborative learning situations has every reason to take into consideration the results of critical studies showing how some student profiles benefit more than others from the usual group work settings (Mayberry, 1998).

Finally, while the scientific café format studied here does not offer the time to go beyond the expression and confrontation of students' conceptions with each other and with the few elements of knowledge provided in the "information desks" of the slide show, it would be promising to design a longer pedagogical sequence on the same SSC. In particular, it would be interesting to include the transition from oral argumentation to a form of "written" conclusions that would open students' reasoning to another temporality, whether it be a classic textual format or other formats such as a mind map, a poster, a movie, etc. Such transition would certainly be both semiotic and sociocognitive, implying a shift to norms differing from the canons of a semi-formal oral. It would become possible to analyze what the students themselves would assume and present as "argumentative results" of their debates, certainly different from the totality of the things said during these discussion. Undoubtedly, such proess would also bring into play other forms of emotions, related to writing. How to account for the collective exploration of a controversial issue, which, by definition, does not lead to a consensus? Such work would require further development of empathic skills in order to succeed in reporting, without caricaturing them, the points of view that we do not share, and to make explicit what definitely opposes us: these cognitive bundles consisting of disagreement on the description of the problem, its affective framing, and the desirable alternatives to imagine.

References

Abelson, R., & Schank, R. C. (1977). *Scripts, plans, goals and understanding. An inquiry into human knowledge structures.* New Jersey.

Aikenhead, G. (1992). The integration of STS into science education. *Theory into practice*, 31(1), 27–35.

Ainsworth, S., Gelmini-Hornsby, G., Threapleton, K., Crook, C., O'Malley, C., & Buda, M. (2010). Anonymity in classroom voting and debating. *Learning and Instruction*, 21(3), 365–378.

Allwood, J., Traum, D., Jokinen, K. (2000). Cooperation, Dialogue and Ethics. *International Journal of Human-Computer Studies*, 53, 871–914.

Albe, V. (2009a). L'enseignement de controverses socioscientifiques. *Éducation & didactique*, 3(1), 45–76.

Albe, V. (2009b). *Enseigner des controverses.* Rennes : Presses universitaires de Rennes.

Albe, V. & Simonneaux, L. (2002). L'enseignement des questions scientifiques socialement vives dans l'enseignement agricole : quelles sont les intentions des enseignants ? *Aster*, 34, 131–156.

Albe, V. (2006). Procédés discursifs et rôles sociaux d'élèves en groupes de discussion sur une controverse socio-scientifique. *Revue française de pédagogie*, 157, 103–118.

Alexander, R. J. (2017). *Towards Dialogic Teaching: rethinking classroom talk* (5e éd.). Dialogos.

Amossy, R. (2006). *L'argumentation dans le discours.* Armand Colin.

Andriessen, J., Baker, M., & Suthers, D. (2003). *Arguing to learn: Confronting cognitions in computer-supported collaborative learning environments.* Springer Netherlands.

Andriessen, J., Pardijs, M., Baker, M. (2013). Getting on and Getting Along: Tension in the Developement of Collaborations. Dans M. Baker, S. Järvelä, & J. Andriessen, (dir.) *Affective Learning Together* (p. 205–228). Routledge.

Andriessen, J. (2006). Arguing to learn. Dans R. K. Sawyer (dir.), *The Cambridge handbook of the learning sciences* (2e éd., p. 443–460). Cambridge University Press.

Anscombre, J. C., & Ducrot, O. (1997 [1981]), *L'argumentation dans la langue.* Éditions Mardaga.

Aristote. 2007. *Rhétorique.* Translated by Pierre Chiron. Paris : Garnier-Flammarion.

Arppe, A., Gilquin G., Glynn, D., Hilpert, M. et Zeschel, A. (2010). Cognitive Corpus Linguistics: five points of debate on current theory and methodology. *Corpora*, 5, 1–27.

Asterhan, C. S. C. (2013). Epistemic and interpersonal dimensions of peer argumentation. Dans M. Baker, S. Järvelä, & J. Andriessen, (dir.) *Affective Learning Together* (p. 251–271). Routledge.

Audigier, F. (1999). *L'éducation à la citoyenneté*. Lyon : INRP.

Baker, M., Järvelä, S., & Andriessen, J. (Dir.) (2013). *Affective Learning Together: Social and Emotional Dimensions of Collaborative Learning.* Routledge.

Baker, M., Quignard, M., Lund, K., & van Amelsvoort, M. (2002). Designing a computer-supported collaborative learning situation for broadening and deepening understanding of the space of debate. *Proceedings of the 5th International Conference of the International Society for the Study of Argumentation.* Amsterdam. 55-61.

Bakhtin, M. (1981). *Discourse in the Novel* [1934/35] (trad. C. Emerson & M. J. Holquist, éd. M. J. Holquist). The Dialogic Imagination. 258-422.

Brown, P., & Levinson, S. C. (1988). *Politeness.* Cambridge University Press

Brundtland, G. H. (1987). *Our Common Future.* United Nations.

Buty, C., & Plantin, C. (2008). L'argumentation à l'épreuve de l'enseignement des sciences et vice-versa. *Argumenter en classe de sciences. Du débat à l'apprentissage* (p. 17-42). Lyon : INRP.

Buty, C. & Plantin, C. (2008). *Argumenter en classe de sciences : du débat à l'apprentissage.* Lyon: INRP.

Bybee, R. W. (1987). Science education and the science-technology-society (S-T-S) theme. *Science Education*, 71(5), 667-683.

Callon, M., Lascoumes, P., & Barthe, Y. (2001). *Agir dans un monde incertain : essai sur la démocratie technique.* Paris : Éditions du Seuil.

Carrington, B., & Troyna, B. (1988). *Children and controversial issues: strategies for the early and middle years of schooling.* Psychology Press.

Chevallard, Y. (1992). Concepts fondamentaux de la didactique : perspectives apportées par une approche anthropologique. *Recherches en didactique des mathématiques*, 12(1), 73-112.

Clarke, P. (1992). Teaching controversial issues. *Green Teacher*, (31), 29-32.

Cornelius, L. L., & Herrenkohl, L. R. (2004). Power in the classroom: How the classroom environment shapes students' relationships with each other and with concepts. *Cognition and Instruction*, 22(4), 467-498.

Cosnier, J. (1994). *Psychologie des émotions et des sentiments.* Paris: Retz.

Damasio, A. R. (1995). *L'Erreur de Descartes. Les raisons des émotions.* Paris: Odile Jacob.

Dearden, P. (1981). Public participation and scenic quality analysis. *Landscape Planning*, 8(1), 3-19.

Denis, A., Quignard, M., Fréard, D., Détienne, F., Baker, M. (2012). Détection de conflits dans les communautés épistémiques en ligne. Proceedings of TALN 2012 (2012, Grenoble, France). In G. Antoniadis, H. Blanchon, G. Sérasset (dir.), *Actes de la conférence conjointe JEP-TALN-RECITAL* (p. 351-358). Grenoble: GETALP-LIG.

Désautels, J., & Larochelle, M. (1998). The epistemology of students: The « thingified » nature of scientific knowledge. *International handbook of science education*, 1998, 115-126.

Dewhurst, D. W. (1992). The teaching of controversial issues. *Journal of Philosophy of Education*, 26(2), 153-163.

Dillenbourg, P., & Jermann, P. (2010). Technology for classroom orchestration. In M. S. Khine & I. M. Saleh, (dir.). *New science of learning: Cognition, computers and collaboration in education* (p. 525-552). Springer.

Doise, W., & Mugny, G. (1981). *Le développement social de l'intelligence*. InterEditions.

Douaire, J. (2004). *Argumentation et disciplines scolaires*. Saint-Fons: INRP.

Doury, M. (2004). La classification des arguments dans les discours ordinaires. *Langages*, (2), 59-73.

Driver, R., Leach, J., Millar, R., & Scott, P. (1996). *Young Peoples's Images of Science*. Philadelphia: Open University Press.

Driver, R., Newton, P., & Osborne, J. (2000). Establishing the norms of scientific argumentation in classrooms. *Science Education*, 84(3), 287-312.

Duschl, R. (2000). Making the nature of science explicit. In R. Millar, J. Leach & J. Osborne (dir.), *Improving science education: The contribution of research* (p. 187-206).

Erduran, S., & Jiménez-Aleixandre, M. P. (2007). *Argumentation in Science Education: Recent Developments and Future Directions*. Dordrecht: Springer.

Fernández, M., Wegerif, R., Mercer, N., & Rojas-Drummond, S. (2002). Re-conceptualizing scaffolding and the zone of proximal development in the context of symmetrical collaborative learning. *Journal of Classroom Interaction*, 36(2/1), 40-54.

Goffman, E. (1974). *Les rites d'interaction*. Paris: Editions de Minuit.

Grize, J. B. (1990). *Logique et langage*. Ophrys.

Grize, J. B. (1996). *Logique naturelle et communications*. Paris: Presses universitaires de France.

Guerrini J-C. (2015). *Les valeurs dans l'argumentation. Structures axiologiques et dimension axiologique des disputes*, doctoral dissertation. Université Lyon 2, Lyon.

Hamblin, C. L. (1970). *Fallacies*. Methuen.

Hekmat, I, Micheli, R., Rabatel, A. (dir.) (2013). Modes de sémiotisation et fonctions argumentatives des émotions. *Semen*, 35.

Henderson, J., & Lally, V. (1988). Problem solving and controversial issues in biotechnology. *Journal of Biological Education*, 22(2), 144-150.

Järvenoja, H. & Järvelä, S. (2013). Regulating Emotions Together for Motivated Collaboration. In M. Baker, S. Järvelä, J. Andriessen (dir.), *Affective Learning Together* (p. 162-181). Routledge.

Jimenez–Aleixandre, M. P. (2006). Les personnes peuvent–elles agir sur la réalité ? La théorie critique et la marée noire du Prestige. In A. Legardez & L. Simonneaux (dir.), *L'école à l'épreuve de l'actualité : enseigner les questions vives*, (p. 105–118). ESF Éditeur.

Joshua, S. & Dupin, J.-J. (1993). *Introduction à la didactique des mathématiques et des sciences*. Paris: PUF.

Kelly, T. E. (1986). Discussing controversial issues: Four perspectives on the teacher's role. *Theory and Research in Social Education*, 14(2), 113–138.

Latour, B. & Woolgar, S. (1979). *Laboratory Life: The Social Construction of Scientific Facts*. Beverly Hills, Sage Publications.

Legardez, A. (2006). Enseigner des questions socialement vives : quelques points de repères. In A. Legardez & L. Simonneaux (dir.), *L'école à l'épreuve de l'actualité* (p. 19–32). Paris: ESF.

Lewis, J. & Leach, J. (2006). Discussion of socio–scientific issues: The role of science knowledge. *International Journal of Science Education*, 28(11), 1267–1287.

Mayberry, M. (1998). Reproductive and resistant pedagogies: The comparative roles of collaborative learning and feminist pedagogy in science education. *Journal of Research in Science Teaching*, 35, 443–459.

Menary, R., (2010). Introduction to the special issue on 4E cognition. *Phenomenology and the Cognitive Sciences*, 9, 459–463.

Mercer, N. (1996). The quality of talk in children's collaborative activity in the classroom. *Learning and instruction*, 6(4), 359–377.

Mercer, N. & Littleton, K. (2007). *Dialogue and the development of children's thinking: a sociocultural approach*. Psychology Press.

Mercer, N. & Sams, C. (2006). Teaching Children How to Use Language to Solve Maths Problems. *Language and Education*, 20(6), 507–528.

Meuffels, B. & van Eemeren, F. H. (2002). Ordinary arguers' judgments on *ad hominem* fallacies. *Advances in pragma–dialectics*, 45–64.

Micheli, R. (2010). *L'émotion augmentée : l'abolition de la peine de mort dans le débat parlementaire français*. Cerf.

Micheli, R. (2013). Esquisse d'une typologie des différents modes de sémiotisation verbale de l'émotion, *Semen*, 35, 17–39.

Ministère de l'éducation nationale, de l'enseignement supérieur et de la recherche (MENESR). (2015). *Le socle commun de connaissances, de compétences et de culture*. Décret du 31 mars 2015, JO du 2 avril 2015.

Ministère de l'éducation nationale, de l'enseignement supérieur et de la recherche (MENESR). (2006). *Le socle commun de connaissances et de compétences, Tout ce qu'il est indispensable de maîtriser à la fin de la scolarité obligatoire*. Décret du 11 juillet. Ed. CNDP.

Mortimer, E., & Scott, P. (2003). *Meaning making in secondary science classrooms*. Open University Press.

Muller–Mirza, N. (2008). Préface. Buty, C. & Plantin, C. *Argumenter en classe de sciences*, INRP, 7–16.

Mullins, Deiglmayr, Spada. (2013). Motivation and emotion shaping knowledge co-construction. In M. Baker, S. Järvelä, & J. Andriessen, (dir.) *Affective Learning Together* (p. 139-160). Routledge.

Muntigl, P. & Turnbull, W. (1998). Conversational structure and facework in arguing. *Journal of Pragmatics*, 29(3), 225-256.

Newton, P., Driver, R., & Osborne, J. (1999). The place of argumentation in the pedagogy of school science. *International Journal of Science Education*, 21(5), 553-576.

Ngo, T. T. H. (2011). *Argumentation et didactique du français langue étrangère pour un public vietnamien*, doctoral dissertation. Université Lyon 2, Lyon, France.

Oetzel, J., Ting-Toomey, S., Masumoto, T., Yokochi, Y., Pan, X., Takai, J., & Wilcox, R. (2001). Face and facework in conflict: A cross-cultural comparison of China, Germany, Japan, and the United States. *Communication Monographs*, 68(3), 235-258.

Osborne, J., Erduran, S., & Simon, S. (2004). Enhancing the quality of argumentation in school science. *Journal of Research in Science Teaching*, 41(10), 994-1020.

Oulton, C., Dillon, J., & Grace, M. M. (2004). Reconceptualizing the teaching of controversial issues. *International Journal of Science Education*, 26(4), 411-423.

Perelman, C., Olbrechts-Tyteca, L. (1988) [1958]. *Traité de l'argumentation : la nouvelle rhétorique*. Université de Bruxelles.

Plantin, C. (2018). *Dictionary of argumentation: an introduction to argumentation studies*. College Publications.

Plantin, C. (2015). Les séquences discursives émotionnées : définition et application à des données tirées de la base CLAPI. *SHS Web of Conferences, 3e congrès mondial de linguistique française*, 1, 629 642.

Plantin, C. (2011). *Les bonnes raisons des émotions : principes et méthode pour l'étude du discours « émotionné »*. Peter Lang.

Plantin, C. (1996). *L'argumentation*. Paris : Seuil.

Plantin, C., Doury, M., & Traverso, V. (2000). *Les émotions dans les interactions.* Presses universitaires de Lyon.

Polo, C., Lagrange-Lanaspre, S. (2019), Metaphorical Reasoning Together: Embodied Conceptualization in a Community of Philosophical Inquiry. In K. Lund, G. Niccolai, E. Lavoué, C. Hmelo-Silver, G. Gweon, M. Baker (Eds), *A Wide Lens: Combining Embodied, Enactive, Extended, and Embedded Learning in Collaborative Settings: 13th International Conference on Computer Supported Collaborative Learning*, (1) 424-43.

Polo, C. (2018). Can experiencing non-linear scripting affect argumentative emotional positioning? Symposium Argumentation, emotion and scripting: learning sciences and interactive narrative design. EARLI SIG 20-26 Conference "Argumentation and Inquiry as Venues for Civic Education". Jérusalem, October 8-12.

Polo, C., Plantin, C. Lund, K., Niccolai, G. P. (2017a). Group Emotions in Collective Reasoning: a Model. *Argumentation*, 31(2), 301-329.

Polo, C., Plantin, C. Lund, K., Niccolai, G. P (2017b). Emotional Positioning as a Cognitive Resource for Arguing. *Pragmatics and Society*, 8(3), 323-354.

Polo, C., K. Lund, C. Plantin, G. P. Niccolai (2016b). Group Emotions: The Social and Cognitive Functions of Emotions in Argumentation. *International Journal of Computer-Supported Collaborative Learning*, 11(2), 123-156.

Polo, C., Plantin, C. Lund, K., Niccolai, G. P. (2016a), Words to Reason and Argue about Drinking Water Management: Comparing Debates from Mexican, US and French Schools. In D. Mohammed & M. Lewínski (dir.), *Argumentation and Reasoned Action: Proceedings of the 1st European Conference on Argumentation*, Lisbon, 2015 (Vol. II, 821-838). Londres : College Publications.

Polo, C., Plantin, C., Lund, K., Niccolai, G. (2013b). Quand construire une position émotionnelle, c'est choisir une conclusion argumentative : le cas d'un café-débat sur l'eau potable au Mexique. *Semen*, 35, 41-63.

Polo, C., Plantin, C., Lund, K., Niccolai, G. P. (2013a). Cohering Without Converging: Students' Use of Doxa, Norms and Values while Debating about SSI (Mexico, USA, France). *Proceedings of the Conference of the European Science Education Research Association* (ESERA). Nicosia, Cyprus, September 2012, 1387-1399.

Pomerantz, A. (1984). Agreeing and disagreeing with assessments: Some features of preferred/dispreferred turn shapes. In J. Maxwell Atkinson & John Heritage (dir.) *Structures of social action: Studies in conversation analysis* (p. 57-101). Cambridge: Cambridge University Press.

Roschelle, J. & Teasley, S. D. (1995). The construction of shared knowledge in collaborative problem solving. Proceedings of the *Conference on Computer-supported collaborative learning*, 69-97.

Sadler, T. D. & Fowler, S. R. (2006). A threshold model of content knowledge transfer for socioscientific argumentation. *Science Education*, 90(6), 986-1004.

Sadler, T. D., & Zeidler, D. L. (2005). The significance of content knowledge for informal reasoning regarding socioscientific issues: Applying genetics knowledge to genetic engineering issues. *Science Education*, 89(1), 71-93.

Sandoval, W. A. (2005). Understanding students' practical epistemologies and their influence on learning through inquiry. *Science Education*, 89(4), 634-656.

Schwarz, B., & Glassner, A. (2007). The role of floor control and of ontology in argumentative activities with discussion-based tools. *International Journal of Computer-Supported Collaborative Learning*, 2(4), 449-478.

Sensevy, G. (2007). Des catégories pour décrire et comprendre l'action didactique. In G. Sensevy & A. Mercier, *Agir ensemble: Eléments de théorisation de l'action conjointe du professeur et des élèves* (p. 13-49). Rennes : Presses Universitaires de Rennes.

Simon, J. P. & Tozzi, M. (dir.). (2017). *Paroles de philosophes en herbe*. Grenoble: UGA Éditions.

Simonneaux, L. (2006). Quel enjeu éducatif pour les questions biotechnologiques ? In A. Legardez & L. Simonneaux, *L'école à l'épreuve de l'actualité : enseigner des questions vives* (p. 33-61). Issy-les-Moulineaux : ESF.

Simonneaux, L. (2003). L'argumentation dans les débats en classe sur une technoscience controversée. *Aster*, 37, Interactions langagières 1, 189-214.

Simonneaux, L. (1995). *Approche didactique et muséologique des biotechnologies de la reproduction bovine*, doctoral dissertation. Université Claude Bernard, Lyon, France.

Simonneaux, L. & Bourdon, A. (1998). Antigen, antibody, antibiotics... What did you say that was? In H. Bayrhuber & H. Brinkman, What-Why-How? *Research in Didaktik of Biology* (p. 233_242). Kiel: IPN.

Simonneaux, L., & Simonneaux, J. (2005). Argumentation sur des questions sociocientifiques. *Didaskalia*, 27, 79-108.

Sins, P. & Karlgren, K. (2013). Identifying and Overcoming Tension in Interdisciplinary Teamwork in Profesional Development. In M. Baker, S. Järvelä, & J. Andriessen, (dir.), *Affective Learning Together* (p. 185-203). Routledge.

Stahl, G. (2006). *Group cognition*. Cambridge: MIT Press.

Tiberghien, A. (2009). Préface. In V. Albe, *Enseigner des controverses* (p. 3-10). Rennes: Presses universitaires de Rennes.

Toulmin, S. E. (2003 [1958]). *The uses of argument*. Cambridge University Press.

Vignaux, G. (1976). *L'Argumentation. Essai d'une logique discursive*. Genève: Droz.

Van Eemeren, F. H. & Grootendorst, R. (2004). *A systematic theory of argumentation: The pragma-dialectical approach*. Cambridge University Press.

Walton, D. N. (1992). *The place of emotion in argument*. Pennsylvania State University Press.

Wegerif, R., Littleton, K., Dawes, L., Mercer, N., & Rowe, D. (2004). Widening access to educational opportunities through teaching children how to reason together. *Westminster Studies in Education*, 27(2), 143-156.

Wegerif, R., & Mercer, N. (1997). A dialogical framework for researching peer talk. *Language and Education Library*, 12, 49-64.

Weinberger, A., Reiserer, M., Ertl, B., Fischer, F., & Mandl, H. (2005). Facilitating collaborative knowledge construction in computer-mediated learning environments with cooperation scripts. In R.

Bromme, F. W. Hesse & H. Spada, *Barriers and biases in computer-mediated knowledge communication*, 15–37.

Zeidler, D. L., Sadler, T. D., Simmons, M. L., & Howes, E. V. (2005). Beyond STS: A research-based framework for socioscientific issues education. *Science Education*, 89(3), 357–377.

Appendix: Conventions used to transcribe students' debates

N LOC utterance	speech turn number, space, first three letters of the pseudonym of the speaker, utterance assigned to him/her at this turn
LOC ((action))	non-verbal turn
<((laughing)) utterance>	comment on the coverbal
:	elongated sound (pasted after the sound)
/ or \	rising or falling intonation, at the end of the word
&	at the end of the turn, indicating that the turn is not really finished, and at the beginning of the turn, indicating the continuation of a previous turn
=	quasi-overlap: very fast change of speaker, at the end of a turn and the beginning of the next one
[	beginning of overlap and possibly] at the end of the overlap
xxx	inaudible segment
(word)	uncertain transcription
°word°	low voice, °° for very low
WORD	increased volume (so there is usually no capitalization)
`	non-standard elision
Be–a–u–ti–ful	Speech rythm 'cut' into syllables
(.)	pause, if measured, in seconds, time indicated (2.3), it can be attributed to a speaker in a turn to speech or to everyone, then placed between speech turns

www.ingramcontent.com/pod-product-compliance
Lightning Source LLC
Chambersburg PA
CBHW070657100426
42735CB00039B/2172